T0185337

Series Editor: **Michael Beaney**

Giusseppina D'Oro
REASONS AND CAUSES
Causalism and Non-Causalism in the Philosophy of Action

Anssi Korhonen
LOGIC AS UNIVERSAL SCIENCE
Russell's Early Logicism and Its Philosophical Context

Sandra Lapointe (*translator*)
Franz Prihonsky
THE NEW ANTI-KANT

Consuelo Preti
THE METAPHYSICAL BASIS OF ETHICS
The Early Philosophical Development of G.E. Moore

Erich Reck (*editor*)
THE HISTORIC TURN IN ANALYTIC PHILOSOPHY

Maria van der Schaar
G.F. STOUT: ON THE PSYCHOLOGICAL ORIGIN OF ANALYTIC PHILOSOPHY

History of Analytic Philosophy
Series Standing Order ISBN 978–0–230–55409–2 (hardcover)
Series Standing Order ISBN 978–0–230–55410–8 (paperback)
(*outside North America only*)

You can receive future titles in this series as they are published by placing a standing order. Please contact your bookseller or, in case of difficulty, write to us at the address below with your name and address, the title of the series and one of the ISBNs quoted above.

Customer Services Department, Macmillan Distribution Ltd, Houndmills, Basingstoke, Hampshire RG21 6XS, England

Frege on Absolute and Relative Truth

An Introduction to the Practice of Interpreting Philosophical Texts

Ulrich Pardey

Ruhr Universität Bochum, Germany

palgrave
macmillan

© Ulrich Pardey 2012

Softcover reprint of the hardcover 1st edition 2013 978-1-137-01222-7

All rights reserved. No reproduction, copy or transmission of this publication may be made without written permission.

No portion of this publication may be reproduced, copied or transmitted save with written permission or in accordance with the provisions of the Copyright, Designs and Patents Act 1988, or under the terms of any licence permitting limited copying issued by the Copyright Licensing Agency, Saffron House, 6–10 Kirby Street, London EC1N 8TS.

Any person who does any unauthorized act in relation to this publication may be liable to criminal prosecution and civil claims for damages.

The author has asserted his right to be identified as the author of this work in accordance with the Copyright, Designs and Patents Act 1988.

First published 2012 by
PALGRAVE MACMILLAN

Palgrave Macmillan in the UK is an imprint of Macmillan Publishers Limited, registered in England, company number 785998, of Houndmills, Basingstoke, Hampshire RG21 6XS.

Palgrave Macmillan in the US is a division of St Martin's Press LLC, 175 Fifth Avenue, New York, NY 10010.

Palgrave Macmillan is the global academic imprint of the above companies and has companies and representatives throughout the world.

Palgrave® and Macmillan® are registered trademarks in the United States, the United Kingdom, Europe and other countries

ISBN 978-1-349-43653-8   ISBN 978-1-137-01223-4 (eBook)
DOI 10.1057/9781137012234

This book is printed on paper suitable for recycling and made from fully managed and sustained forest sources. Logging, pulping and manufacturing processes are expected to conform to the environmental regulations of the country of origin.

A catalogue record for this book is available from the British Library.

A catalog record for this book is available from the Library of Congress.

10 9 8 7 6 5 4 3 2 1
21 20 19 18 17 16 15 14 13 12

To Barbara, Eva and Katja

Contents

Series Editor's Preface

During the first half of the twentieth century analytic philosophy gradually established itself as the dominant tradition in the English-speaking world, and over the last few decades it has taken firm root in many other parts of the world. There has been increasing debate over just what 'analytic philosophy' means, as the movement has ramified into the complex tradition that we know today, but the influence of the concerns, ideas and methods of early analytic philosophy on contemporary thought is indisputable. All this has led to greater self-consciousness among analytic philosophers about the nature and origins of their tradition, and scholarly interest in its historical development and philosophical foundations has blossomed in recent years. The result is that history of analytic philosophy is now recognized as a major field of philosophy in its own right.

The main aim of the series in which the present book appears – the first series of its kind – is to create a venue for work on the history of analytic philosophy, consolidating the area as a major field of philosophy and promoting further research and debate. The 'history of analytic philosophy' is understood broadly, as covering the period from the last three decades of the nineteenth century to the start of the twenty-first century – beginning with the work of Frege, Russell, Moore and Wittgenstein, who are generally regarded as its main founders, and the influences upon them – and going right up to the most recent developments. In allowing the 'history' to extend to the present, the aim is to encourage engagement with contemporary debates in philosophy – for example, in showing how the concerns of early analytic philosophy relate to current concerns. In focusing on analytic philosophy, the aim is not to exclude comparisons with other – earlier or contemporary – traditions, or consideration of figures or themes that some might regard as marginal to the analytic tradition but which also throw light on analytic philosophy. Indeed, a further aim of the series is to deepen our understanding of the broader context in which analytic philosophy developed, by looking, for example, at the roots of analytic philosophy in neo-Kantianism or British idealism, or the connections between analytic philosophy and phenomenology, or discussing the work of philosophers who were important in the development of analytic philosophy but who are now often forgotten.

Although Gottlob Frege (1848–1925) is now widely recognized as one of the key founders of the analytic tradition, his work was appreciated by very few in his lifetime, and it was only through his influence on Russell and Wittgenstein, in particular, that his writings came to be studied in their own right. Even now, though, interpretations of his philosophy are often distorted by later views. Frege's work is rooted in his creation of quantificational logic, which he used to pursue his logicist project, intended to demonstrate that arithmetic can be reduced to logic. In doing so, Frege was led to think through various epistemological issues and to develop new ideas that formed the basis of modern philosophy of language – most famously, the distinction between sense (*Sinn*) and reference (*Bedeutung*), which was explained in an article published in 1892. His logicist project, however, foundered on the paradox that Russell informed him about in 1902 and that now bears Russell's name; and it was left to Russell – with the later help of Whitehead – to pursue it further. Frege abandoned his logicist project, but he continued to write, defending and explaining his key logical ideas.

Between 1918 and 1923, after his retirement from the University of Jena, Frege published a series of three articles under the general title 'Logical Investigations'. The first of these, entitled 'Thought' ('Der Gedanke') has become as well known as his earlier essay on sense and reference. Frege here explains his conception of thought as the objective content of thinking, its connection with truth, and the tripartite distinction he draws between physical things, mental ideas and thoughts, the latter seen as inhabiting a 'third realm'. In discussing the relationship between thought and truth at the beginning of the article, and in particular, in the third and fourth paragraphs, Frege criticizes the correspondence theory of truth and argues that truth is indefinable. It is on Frege's arguments in these two paragraphs that Ulrich Pardey focuses in the present book. As Pardey points out, most commentators have read these paragraphs from a perspective informed by the subsequent work of Tarski. In a paper first published in Polish in 1933 (translated into German in 1935 and English in 1956), Tarski argued that truth *is* definable, and his theory of truth has generally been seen as a type of correspondence theory. As a result, Frege's arguments have been assumed to be fallacious, in one way or another. Pardey sets out to show that these arguments are not fallacious, and by distinguishing between absolute truth and relative truth, he defends Frege's view of truth. In doing so, he offers the most careful and convincing reading of Frege's arguments to have appeared in the philosophical literature to date.

In offering this reading, Pardey's book is also intended as an introduction to the practice of interpreting philosophical texts, and in the final chapter he discusses some of the methodological issues involved with this practice. He lays down six rules that an interpretation should follow, and suggests that analytic philosophers are especially prone to reading inconsistencies into texts wherever they seem to conflict with current views. Such interpretations he calls 'incriminating' interpretations, which he distinguishes from 'apologetic' interpretations, exemplified by his own account. So as well as defending Frege's view of truth, the book provides an excellent case study in how to do good work in history of analytic philosophy.

Michael Beaney
May 2012

Preface

This book has two objectives: to be a contribution to the understanding of Frege's theory of truth, and to be an introduction to the practice of interpreting philosophical texts.

*

Its contribution to Frege's theory of truth is basically restricted to an interpretation of the third and fourth paragraphs in Frege's essay 'Der Gedanke' and of a parallel passage in Frege's posthumously published fragment 'Logik'. These paragraphs are well known and notorious for their criticism of the correspondence theory of truth, as well as for Frege's argument for the indefinability of truth, and they are almost invariably cited as classic examples of elementary and easily detectable fallacies. In this book I propose a completely new interpretation of Frege's text, against the standard interpretation and the general consensus among such renowned Frege scholars as Dummett, Künne, Soames and Stuhlmann-Laeisz. I aim to show in Chapters 1–12 that Frege's argumentation is *internally* consistent and can be shown to hold up well against previous objections.

My interpretation centers around Frege's distinction between absolute and relative truth. Tarski declared the concept of absolute truth to be obsolete on the grounds that, as the liar antinomy showed, it was inconsistent or meaningless. In defense of Frege, however, I attempt to show in Chapters 14–15 that neither the liar antinomy nor Tarski's account of truth would necessitate abandoning Frege's absolute concept of truth (for Chapter 13 see below). In short: I aim to show in Chapters 14–15 that Frege's argumentation is *externally* consistent as well.

The success of my proposal of a new interpretation will depend on whether I am able to provide a detailed justification for it and to provide compelling reasons to reject previous alternative interpretations. That is why my interpretation needs to be much more detailed than its established competitors. It will not suffice to present a succinct paraphrase of Frege's text, hoping that every reader will see how the interpretation matches the interpreted text. Instead, I will have to show in detail – and against the consensus among Frege scholars – how previous interpretations failed to do justice to Frege's text and why my new interpretation more accurately renders Frege's original argumentation.

This book is intended as a *contribution* to Frege's theory of truth; however, it does not attempt to be a comprehensive account of this theory, let alone of Frege's philosophy as a whole. Trying to give an account of Frege's philosophy as a whole means letting go of the aim of being precise in the details, while pursuing precision in the details means dispensing with a comprehensive overall account. While I do not deny that there are intermediate positions between these two extremes, I have decided in favor of the second of them. For in the light of the persistent misinterpretations of the text discussed here, Frege scholarship would make a considerable step forward if I succeeded in providing a convincing refutation of these misinterpretations as well as a proof of the consistency of Frege's arguments. In a second step, others as well could draw further consequences from my novel interpretation for the reception of Frege's philosophy as a whole.

My contribution to Frege's theory of truth is essentially limited to an *immanent* interpretation of the indicated text passages. This is why I will not address Frege's disputes with his contemporaries. I will, however, address Tarski, due to the fact that I have looked for an explanation of the indicated misinterpretations. If we carefully read those of Frege's texts that are the subject of the interpretations in this book, then we have to ask how those misinterpretations could have been possible. The only answer that I can come up with is this: Frege's texts have been read under the influence of Tarski's theory of truth, which has led to their being misunderstood.

* *

The second objective of this book, to provide an exemplar of the interpretation of philosophical texts, also requires great thoroughness and detail. Many interpretations of philosophical texts leave the reader confused about how the interpretation is based on the original text. The interpreters mention the text, perhaps quote it, and then formulate their own paraphrases as an interpretation of the text. But how they got from the text to the paraphrases remains a mystery. For students of philosophy, striving to learn the art of interpretation, it is very important, however, to recognize the *road* from the text to its interpretation and to learn to walk that road independently, if they wish to genuinely engage in philosophizing. This is why in Chapter 13 I have added a description of some of the stations along the road by which I arrived at my new interpretation.

Students can learn how to interpret philosophical texts by critically studying the controversial interpretations of Frege's aforementioned text that are discussed or proposed in this book by examining these interpretations step by step, and by comparing both interpretations time and again with the text itself. This examination of different interpretations forces students to read the original text slowly, over and over again as if anew, and thoroughly. Such thorough reading, which is a precondition of all interpretation, can thus be trained or practiced at a deeper level with the help of this book. Hence I believe that this book may well serve as a useful introduction to the practice of interpreting argumentative philosophical texts.

This introduction differs from other introductions to the art of interpreting especially in two respects: First, it does not present a theory of interpretation; hence it is not a contribution to hermeneutics but an introduction to the *practice* of interpreting. In my view we learn the art of interpreting better by following a good practical example than by just studying a theory. This is why I present this book as a 'paradigm example' of an interpretation. While in Chapters 13 and 17 I shall briefly reflect upon the practice of interpreting here demonstrated, these chapters do not develop or discuss a theory of interpretation but merely contain some advice for the practice of interpreting. In short: This is an introduction to the *practice*, not the theory, of interpreting.

Second, the kind of interpretation practiced in this book is not uniformly applicable to *all* philosophical texts. Given that there are different kinds of philosophical texts and that interpretations should conform to their respective types of text, I believe that different texts require different ways of interpreting as well. In this book I interpret a text by Frege, and though the same kind of interpretation can be easily transferred to texts containing similar kinds of arguments – such as texts by Aristotle, Thomas Aquinas, Hume, Kant, Bolzano, Quine or Strawson – it may be less suitable for other texts.

* * *

There are at least five reasons why the third and fourth paragraphs of Frege's essay 'Der Gedanke' and the parallel passage from 'Logik' are very well suited to serve as a textual basis for students of philosophy to learn how to arrive at an accurate interpretation.

The first is just that the text is short enough to make reading it so painstakingly that no comma or dot on an 'i' goes without scrutiny is still acceptable and manageable for the student.

The second is that the text is about one of the central concepts of theoretical philosophy, namely truth; hence, it commands a general interest in its subject matter. It is important here to note that Frege's argumentation is not merely of historical interest, but also constitutes an important contribution to our contemporary discussion of the concept of truth. And, of course, after 'Über Sinn und Bedeutung', 'Der Gedanke' is probably the most famous of Frege's essays; it is mandatory reading for philosophy majors at colleges and universities throughout the world, and is even quoted from in some textbooks for philosophy classes at the high-school level.

The third reason that this text is so suitable for teaching the craft of philosophical interpretation is that it is no longer immediately comprehensible for us today, but in fact requires much interpretation. However, with thorough and persistent reading it can be understood on its own; it requires no special expertise other than the basic logical skills needed for distinguishing between one-place and two-place predicates. Even Frege's critique of psychologism, which plays a crucial role in his argumentation, is sufficiently laid out in the text, albeit in the first paragraph. In this sense the text offers everything that we need to know in order to understand it, and we are in principle able to arrive at our own interpretation by means of a thorough reading of it alone. This represents a great advantage for students when compared to other texts that can be understood only with the help of substantial specialized knowledge.

The fourth reason the text is good for this purpose is that interpretations by scholars such as Dummett, Künne, Soames and so on depict the text as a collection of blunt errors, confronting the reader with a choice between either regarding these scholars' interpretations as correct and accepting that Frege's argumentation is full of mistakes, or regarding Frege's argumentation as not so completely off and seeing those interpretations as blunt misinterpretations. Grappling with this controversy – should we believe the author or those scholars? – can be extremely stimulating for readers, forcing them to examine Frege's text, previous interpretations, and my new interpretation, on their own. And, of course, those who find that they cannot fully agree with any of the interpretations on offer are challenged to develop one of their own. In doing so, they can indirectly learn from the interpretations at hand by straightforwardly aiming at an understanding of the text that deviates from the foregoing ones. In this way even misinterpretations can assist in the development of a better, more accurate interpretation.

Finally, the fifth reason this text works so well for teaching philosophical interpretation is that it is internally consistent, which is something that cannot be said about all philosophical texts. A generous scholar should always go about interpreting a text by initially presuming that the argumentation within the text can be reconstructed in a self-consistent way; unfortunately, this presumption often turns out to be false. If in interpreting Frege's text we consistently hold on to this presumption then we'll see (assuming that my interpretation is correct) that the alleged logical errors in his arguments really are a result of misinterpretations. But this is really one of the most important experiences for those who wish to learn the art of interpretation: the experience, namely, of seeing how a text that may initially seem inconsistent (and be rejected on that basis) can gradually become clearer and lose its appearance of inconsistency under more intensive scrutiny. It is only after this experience that the practice of interpretation becomes a truly worthwhile activity.

* * * *

It is in this regard that the interpretation proposed in this book succeeds in serving both objectives: it serves as a useful exemplar of the interpretation of philosophical texts, and as a contribution to the understanding of Frege's theory of truth.

Ulrich Pardey, Ruhr-Universität Bochum, Germany

Acknowledgments

At this point I would like to mention the people who have contributed to the making of this book. For the period of one academic semester in the summer of 2000, the participants in my seminar on Frege's essay 'Der Gedanke' assisted me in thinking through a difficult text of only 1½ pages, broke their heads over it, and were made to pull their hair out by the combination of the text's intransigence, our thick-wittedness, and the nature of the predominant interpretations in the secondary literature.

My colleagues Friedrich Dudda, Tania Eden, and Hans-Ulrich Hoche, as well as the participants of the Colloquium on Logic and Philosophy of Language at Ruhr-Universität Bochum (in particular Michael Knoop) critically read versions of parts of this book and discussed them with me. Göran Sundholm and Kai Wehmeier provided me with the opportunity to present my interpretation of Frege's critique of the correspondence theory at a conference on Frege in Leiden, Netherlands, in August 2001; I am particularly grateful for Kai Wehmeier's comments both on that occasion and subsequently, as they contained many helpful suggestions.

My assistants Ilke Aydin, Benedikt Fait, and Daniela Zumpf critically read several earlier versions of this book, double-checked the quotes, and repeatedly discussed with me suggested improvements. My sincere thanks to all of them!

Dorothea Lotter, with assistance from Andrew Cross, not only substantially contributed to the translation of the original German version of this manuscript into English, but in the process frequently raised important critical questions and offered very helpful suggestions, thereby also substantially contributing to improvements in my argumentation. My special thanks to both of them for their translation and suggestions!

*

Further I would like to thank *Blackwell* Publishing Ltd., *Felix Meiner* Verlag and *Georg Olms* Verlag for granting permission to quote extensively from Frege's essays published in the following books:

G. Frege (1967) *Kleine Schriften*, ed. I. Angelelli, 2nd edn (Hildesheim: Georg Olms, 1990).

G. Frege (1969) *Nachgelassene Schriften,* ed. H. Hermes, F. Kambartel, F. Kaulbach, 2nd edn (Hamburg: Felix Meiner, 1983).

G. Frege (1979) *Posthumous Writings*, trans. P. Long and R. White, ed. H. Hermes, F. Kambartel, F. Kaulbach (Oxford: Basil Blackwell).

G. Frege (1984) *Collected Papers on Mathematics, Logic, and Philosophy*, ed. B. McGuinness (Oxford: Basil Blackwell).

Finally I would like to thank *Mentis* Verlag for granting permission to publish this – in many respects modified and extended – English version of my book *Freges Kritik an der Korrespondenztheorie der Wahrheit*.

* *

I am very glad that this book appears in the Palgrave Macmillan's History of Analytic Philosophy series and I would like to thank Michael Beaney, the series editor, for admitting this book to this series and Priyanka Gibbons, the Commissioning Editor for Philosophy and Psychology at Palgrave Macmillan, for her help and advice. Ranjan Chaudhuri, who did the copy editing, improved the text and so I would like to thank him, too.

Texts and Translations

My interpretation of Frege's texts is essentially restricted to the third and fourth paragraphs in 'Der Gedanke' and to a parallel passage in 'Logik'.

In section 3.1 I quote the text of the third and fourth paragraphs of 'Der Gedanke' in full. I have structured the text into sub-paragraphs (i)–(x) and added headings as well as insertions. Thus, wherever one of the parenthesized numbers (i), (ii), ..., (x) appears after a Frege quote, this indicates the corresponding sub-paragraph in the text.

In section 7.1 I quote the text of the parallel passage in 'Logik'. There I make a comparison of the two parallel passages in 'Der Gedanke' and in 'Logik', and for this purpose I have added numbers to the sentences of the third and fourth arguments: (G1)–(G12) in 'Der Gedanke' and (L1)–(L12) in 'Logik'.

*

The translations of the third and fourth paragraphs in 'Der Gedanke' and the parallel passages in 'Logik' largely follow the translations by Geach/Stoothoff ('Der Gedanke') and Long/White ('Logik'), respectively. However, in some places we (Lotter/Pardey) have found it necessary to slightly alter those translations in order to bring them closer in content or sentence structure to the German originals (without thereby sacrificing the elements of English grammar or style). The appendix at the end contains a survey of Frege's original passages in German, the respective translations of these passages by Geach/Stoothoff and Long/White, the modified translations and some remarks on the most important differences in the translations.

Other German texts by Frege or other authors are either quoted in translations indicated in the bibliography or, whenever no translation is indicated, are directly translated into English from the German original by Dorothea Lotter.

* *

Neither Frege's text in the third and fourth paragraphs of 'Der Gedanke' nor the parallel passages in 'Logik' contain any *emphasized* passages.

This is why the translated quotes in the appendix do not contain any emphasis either. However, when quoting from these passages elsewhere in this book I repeatedly *italicize* certain parts without marking this as *my* emphasis. For, according to the above, it goes without saying that such italicized passages must be my emphasis simply because Frege does not have any emphasized words in those passages.

Of course, there are further quotes of Frege and quotes from other authors. Whenever I have emphasized some words in these quotes I have noted that it is my emphasis.

* * *

All texts of Frege here quoted (apart from *Die Grundlagen der Arithmetik* and *Grundgesetze der Arithmetik, Band I*) are contained in Frege, *Kleine Schriften* (1967), and in Frege, *Nachgelassene Schriften* (1969). Translations of these texts are contained in Frege, *Collected Papers on Mathematics, Logic, and Philosophy* (1984), and in Frege, *Posthumous Writings* (1979) as well as in Beaney, *The Frege Reader* (1997). *The Frege Reader* also contains excerpts of *Die Grundlagen der Arithmetik* and *Grundgesetze der Arithmetik* so that all quoted texts by Frege can be found in it.

Page numbers occurring after a Frege quote refer to the original publication of the German text. These page numbers are indicated in all aforementioned works except for Frege's *Posthumous Writings*. Therefore the page numbers for *Posthumous Writings* are added in parentheses after the original page numbers. Page numbers occurring after quotes of works by other authors refer to the most recent edition mentioned in the bibliography.

I use the author/date referencing system ('Harvard referencing') in a slightly modified version as proposed by Stewart Candlish in his book *The Russell/Bradley Dispute* (p. xix). The reasons for the modified version are as follows. Regarding my book it is important that Frege's essay 'Logik' (composed 1897, first published 1969) is about 20 years prior to his essay 'Der Gedanke' (composed and published 1918), because in Chapter 7 I provide a detailed interpretation of how Frege altered his argumentation from his prior essay 'Logik' to his later essay 'Der Gedanke'. So it would be confusing for the reader to refer to 'Logik' by 'Frege, 1969' and to 'Der Gedanke' by 'Frege, 1918'. Therefore I prefer, as Candlish analogously does, to refer to 'Logik' by 'Frege, 1897' and to 'Der Gedanke' by 'Frege, 1918'. With respect to Kant we have an analogous problem: I intend to show how Kant influenced Frege, but this is hardly possible by using references such as 'Kant, 1999' and 'Frege, 1918'.

Candlish describes his 'non-standard system of citation' as follows:

> in the main text and final bibliography, the citation date shown following the author's name is the date of original publication (or, occasionally of composition). A separate date is shown in the bibliography for the edition cited only where this differs from the original. (Candlish, 2007, p. xix)

List of Logical Symbols

¬	Negation ('not')
∧	Conjunction ('and')
∨	Disjunction ('or')
→	Conditional ('if/then')
↔	Biconditional ('if and only if', 'iff')
∀	Universal quantifier ('for all')
∃	Existential quantifier ('there exists')
=	Identity ('identical')
≠	Non-identity ('not identical')

1
Introduction: In Tarski's Shadow

1.1 Frege's critique of the correspondence theory

1.1.0 In the third paragraph of his essay 'Der Gedanke', Frege delivers a criticism of the correspondence theory of truth in altogether *three* consecutive arguments, culminating in the conclusion that the 'attempt to explain truth as a correspondence' fails. Subsequently he provides *one* more brief argument to show that 'any other attempt to define truth' is doomed to fail as well. Frege here thinks of truth as an *absolute* concept rather than a relational one; that is, a true thought is supposed to be true *in itself* rather than being merely true *for* one or the other person X, true *in* a language Y, or true *of* an object or state of affairs Z.

Frege's essay was first published in 1918 and its first English translation appeared in 1956. Tarski's essay 'The Concept of Truth in Formalized Languages' was first released in Polish in 1933, its German translation appeared in 1935 and its English translation in 1956. In that essay, Tarski appears to provide an extensive demonstration of what Frege had pronounced to be impossible. Tarski not only *defines* the concept of truth, he seems to define it as a *relational* concept in the sense of the *correspondence theory*:

> We should like our definition to do justice to the intuitions which adhere to the *classical Aristotelian conception of truth* [...] If we wished to adapt ourselves to modern philosophical terminology, we could perhaps express this conception by means of the familiar formula: *The truth of a sentence consists in its agreement with (or correspondence to) reality.* (For a theory of truth which is to be based upon the latter formulation the term 'correspondence theory' has been suggested.) (Tarski, 1944, pp. 14–15)

Tarski's remarks in his introduction to *The Concept of Truth in Formalized Languages* are completely in line with this:

> [...] that throughout this work I shall be concerned exclusively with grasping the intentions which are contained in the so-called *classical* conception of truth ('true – corresponding with reality') in contrast, for example, with the *utilitarian* conception ('true – in a certain respect useful'). (Tarski, 1935, p. 153)

This is not only Tarski's own interpretation of his definition; it is how the definition has generally been read.[1] Thus, we have Popper pointing out to us that: 'Tarski's theory of truth [...] can be regarded, from an intuitive point of view, as a simple elucidation of the idea of *correspondence to the facts*' (Popper, 1963, p. 224). And Davidson takes a similar stance when he writes:

> The semantic concept of truth as developed by Tarski deserves to be called a correspondence theory because of the part played by the concept of satisfaction; for clearly what has been done is that the property of being true has been explained [...] in terms of a relation between language and something else. (Davidson, 1969, p. 758)

1.1.1   In his essay 'Truth before Tarski', Hans Sluga describes Carnap's difficult road from Frege to Tarski and along the way raises the question of:

> [the extent to which] his dependence on Frege [...] stood initially in the way of that reception [i.e., of Tarski's theory of truth].
>
> [...] Frege's view of the nature of truth [...] is most easily characterized by saying that he takes the concept of truth to be *the* basic semantic notion. [...] Frege summarized his overall view [...] in the following words:
>
> > What is distinctive about my conception of logic is that I give primacy to the content of the word 'true', and then immediately go on to introduce a thought as that to which the question 'Is it true?' is in principle applicable. So I do not begin with concepts and put them together to form a judgment; I come to the parts of a thought by analyzing the thought.[2]
>
> It is important to be absolutely clear about the radical implications of this view. It implies, first, that there cannot be any formal definition

of truth and, secondly, that there cannot be a theory of truth spelling out, either definitionally or axiomatically, what truth means in terms of other semantic concepts. Frege's view leads, thus, to the conclusion that a Tarski-type semantics must be illegitimate. (Sluga, 1999, pp. 27ff)

Accordingly, if Frege's conception is legitimate, Tarski's is not, while if Tarski's is legitimate, Frege's is not. This is the core of the controversy. The epistemic aspect of the controversy can be summarized as follows. Those who approach the matter from a Fregean perspective, as Carnap initially did, may have serious difficulties thinking that Tarski's theory of truth is feasible. The reverse is also true: whoever has Tarski as a major influence, as most analytic philosophers after Tarski do, will have serious difficulties even following Frege's critique of the correspondence theory, let alone accepting it.

1.1.2 For a long time, Frege's essay was hardly noticed by academic philosophers. When Frege was finally rediscovered after 1960, scholars read 'Der Gedanke' under the influence of Tarski. Thus, Günther Patzig, who published a new German edition of 'Der Gedanke' in 1966 and has contributed strongly to the dissemination of Frege's philosophy in Germany, writes in his introduction to Frege's *Logische Untersuchungen*:

> [It appears that Frege's] thesis that the concept of 'truth' is indefinable [cannot be maintained.] [...] [For as] thorough investigations by Tarski and others have shown in the meantime, the concept of truth [...] can be defined on the basis of a distinction between object- and metalanguage [...]. (Patzig, 1966, p. 22f)

Even more influential has been Dummett's reconstruction of Frege's critique in his 1973 book *Frege: Philosophy of Language*. There, in the chapter 'Can Truth be Defined?', Dummett reconstructs Frege's argumentation in the third paragraph of 'Der Gedanke' as though the main objective of Frege's critique of the correspondence theory were to prove the indefinability of truth. Indeed, directly following Dummett's reconstruction and critique of Frege's argumentation, we read: 'This objection succeeds in showing that Frege's argument does not sustain the strong conclusion that he draws, namely *that truth is absolutely indefinable*' (Dummett, 1973, p. 443; my emphasis). Yet reading the text exclusively in this way means distorting Frege's argumentation beyond all recognition – even though, to be fair, Frege's claim of the indefinability of truth must come across as spectacular and even absurd to anyone who has internalized Tarski's definition of truth. And it appears that for a long time after Tarski, Frege's text was given a reading according to which the text primarily aims at showing that truth is indefinable.

1.1.3   Dummett, Künne, Soames, and Stuhlmann-Laeisz, all of whom are leading representatives of the contemporary reading of Frege's critique of the correspondence theory, presumably have read Frege's essay under the influence of Tarski. None of them has made much sense of Frege's critique of the correspondence theory, and all of their discussions of Frege's arguments eventually lead to the result that not one of those arguments holds up against criticism and that all five of them are invalid due to *fundamental logical errors*. This is the view that prevails to this day.

1.2   Six incredible errors

1.2.0   Frege's arguments are said to contain the following logical errors:[3]

(1) Frege is not aware that a one-place predicate such as 'x is a wife' can be defined by means of a two-place predicate such as 'x is married to y' in this way: 'x is a wife:= there is a man y, such that x is married to y'. (Künne, Soames, Stuhlmann-Laeisz)

(2) Frege does not fully take into account the logical difference between 'Socrates was perfect and wise' and 'Socrates was perfectly wise'. (Stuhlmann-Laeisz)

(3) Frege thinks that the decision as to 'whether p' is different from the decision as to 'whether it is true that p', even though he at the same time thinks of the sentences 'p' and 'it is true that p' as synonymous. (Künne)

(4) Frege uses the term 'decision' equivocally, sometimes in an epistemic sense and sometimes in an ontic one. (Stuhlmann-Laeisz, similarly Dummett)

(5) Frege fails to see the difference between an infinite regress and a circle; apparently, when speaking of the one he actually means the other. (Dummett, Künne, Stuhlmann-Laeisz)

(6) Frege overlooks the fact that a definition of the type 'Aubergine:= Eggplant' is not circular. (Kutschera)

If all of these claims were true, Frege would have made at least six beginner's mistakes in logic, and all of them in the third paragraph of his essay. If Frege's theses do indeed contradict Tarski's results, then it appears that we are justified in holding him guilty of even the dumbest mistakes. Yet readings that ascribe to Frege, arguably the most important logician after Aristotle, no fewer than *six* such mistakes in *one* paragraph take the risk of committing a *reductio ad absurdum* of themselves. At the very least I shall attempt to provide a novel interpretation of the third paragraph together with the related fourth paragraph; this interpretation shall take into account Frege's *own* presuppositions and thereby prove his arguments to be *free of logical mistakes*.

1.2.1   This is a book on Frege's theory of truth, not Tarski's, even though the latter name appears in it quite frequently. However, it is not until Chapter 15 that Tarski's own conception of truth will come under discussion; any mention of Tarski before that Chapter is almost always with reference to his influence on the reception of Frege – reference, that is, to Tarski's *shadow*, rather than to Tarski himself. The reception of Frege has been greatly colored by Tarski's influence, particularly with regard to the following four theses, all of which are almost universally accepted today:

(1) Truth is definable.
(2) Truth is definable in conformity with the intuitions of the correspondence theory (see above).
(3) The primary bearer of truth is the sentence.
(4) The ordinary concept of truth has to be replaced by a multitude of truth concepts.

All four of these theses are in sharp contrast to Frege's conception of truth, which includes the theses that: (1) proper truth is indefinable, (2) the correspondence theory is wrong, (3) the thought is the proper truth bearer, and (4) truth is absolute.

This book constitutes a defense of Frege's theory of truth. It is aimed at challenging the claim, so long and so widely treated as unassailable, that after Tarski Frege's theory of truth is obsolete.

2
The Context: The Question of Truth Bearers

2.1 Psychologism – Frege's major target

In Frege's work on the foundations of logic and mathematics, psychologism is one of his major objects of criticism. Psychologism regards logic as a part of psychology and ideas as truth-bearers. Already in the first paragraph, Frege makes sure to draw a clear line between logic and psychology:

> [I]t falls to logic to discern the laws of truth. [...] From the laws of truth there follow prescriptions about asserting, thinking, judging, inferring. And we may very well speak of laws of thought in this way too. But there is at once a danger here of confusing different things. People may very well interpret the expression 'law of thought' by analogy with 'law of nature' and then have in mind general features of thinking as a mental occurrence. A law of thought in this sense would be a psychological law. And so they might come to believe that logic deals with the mental process of thinking and with the psychological laws in accordance with which this takes place. That would be misunderstanding the task of logic, for truth has not here been given its proper place. [...] In order to avoid any misunderstanding and prevent the blurring of the boundary between psychology and logic, I assign to logic the task of discovering the laws of truth, not the laws of taking things to be true or of thinking. (Frege, 1918a, pp. 58–9)

Establishing a boundary between logic and psychology is also the purpose of Frege's argument that *thoughts*, rather than subjective *ideas*, are the proper bearers of scientific truth – for only then can 'truth [be]

6

given its proper place' (see above). Unlike ideas, Fregean thoughts are not objects of psychological research; this is a consequence of the following major thesis of the entire essay: 'A thought belongs neither to my inner world as an idea, nor yet to the external world, the world of things perceptible by my senses' (Frege, 1918a, p. 75). Frege's objections to the correspondence theory of truth are primarily intended to show that *ideas* cannot serve as bearers of scientific truth. With this Frege seeks to deprive psychologism (with regard to logic) of the very grounds on which it rests.

2.2   Relative and absolute truth

2.2.0   Frege's essay 'Der Gedanke' begins with the words: 'Just as "beautiful" points the ways for aesthetics and "good" for ethics, so do words like "true" for logic' (Frege, 1918a, p. 58). This division of the disciplines of philosophy recalls Kant's three Critiques. It is worthwhile dwelling on Kant for a moment for comparison, for doing so means taking a first step out of Tarski's shadow.

In ethics, Kant distinguishes between the concept of the good in itself and the concept of the merely relative good: many things are good just because they are good for the sake of something else. But there is something that is good in itself, and this is the good will. Kant writes:

> It is impossible to think of anything at all in the world, or indeed even beyond it, that could be considered good without limitation except a *good will*. [...] A good will is not good because of what it effects or accomplishes, because of its fitness to attain some proposed end, but only because of its volition, that is, it is good in itself and, regarded for itself, is to be valued incomparably higher than all that could merely be brought about by it in favor of some inclination and indeed, if you will, of the sum of all inclinations. [...] Usefulness or fruitlessness can neither add anything to this worth nor take anything away from it. (Kant, 1785, pp. 49–50)

See also the following passage:

> That is *good* which pleases by means of reason alone, through the mere concept. We call something *good for something* (the useful) that pleases only as a means; however, another thing is called *good in itself* that pleases for itself. (Kant, 1790, pp. 92–3)

Elsewhere Kant writes:

> [...] the [moral] law determines the will *immediately*, the action in conformity with it is *in itself good*, and a will whose maxim always conforms with this law is *good absolutely, good in every respect* and the *supreme condition of all good*. (Kant, 1788, p. 190)

While according to the passage quoted first, the will is good *in itself*, this third passage adds that it is also good *in every respect*, that is, perfectly good. This is just like Frege's notion that a thought is true in every respect, that is, perfectly true (see Chapter 5). There is, then, a certain analogy between Kant's conception of the good will, which is good *in itself* and *perfectly* good, and Frege's conception of the true thought, which is true *in itself* and *perfectly* true. The distinction between the good *in itself* and the good *for* something else is fundamental for Kant; it is also strongly emphasized in Neo-Kantianism, with which Frege was familiar. As Nikolai Hartmann points out:

> The useful is nothing like the good in the sense of ethics. But language lends itself to conceptual confusions. For we also say that something is 'good for something'. Yet, this is not the ethical meaning of 'good'. This meaning can be grasped only once we start asking 'for what' something is good. If we can trace the 'for what' back to that which is no longer merely good for something else, but good in itself, then we have found the good in that other sense that includes its ethical dimension. The ethical good is good in itself. (Hartmann, 1926, p. 88)

Both concepts of the good can be construed as one-place predicates:

'x is (relatively) good', and 'x is good (in itself)'.

Now if we define a one-place predicate such as 'x is good' by means of a two-place predicate such as 'x is good for y', then we obtain the concept of the relative good:

x is (relatively) good:= There is a y (distinct from x) such that: x is good for y.

But, obviously, the good in itself – whether or not it exists – cannot be properly defined in this way.

2.2.1 Analogously, we can define the (one-place) predicate 'x is a wife' by means of a two-place predicate:

x is a wife:= There is a man y such that: x is married to y.

Since every wife must be the wife *of* somebody, this definition of a 'relative' wife suffices. There is no need to define the concept of a wife in herself; in fact, such a concept would be self-contradictory.

Thus, there is only *one* one-place concept of a wife (being a wife is a relational property). But for Kant there are *two* distinct one-place concepts of the good: Only the relative good is a relational property, not the good in itself or the absolute good.

Just as Kant sticks to the concept 'x is *good* in itself', Frege sticks to the concept 'x is *true* in itself'. For Frege, truth is not what is true *for* somebody, or true *of* an object, or true *in* a language, but it is what is true in itself, or in other words: *absolute* truth.

If we are under the influence of Tarski, we may regard truth in itself as just as superfluous and self-contradictory as a wife in herself. Yet we cannot hope to gain a proper understanding of Frege's arguments unless we at least provisionally allow there to be a distinction between truth in itself or absolute truth on the one hand, and relative truth on the other.

2.3 Excursion: relativizations of truth

Concerning the relativization of truth we may distinguish three concepts of relativization.

(1) Truth as a relational concept

The concept of truth is relativized when we conceive of truth as a *relational* concept – as, for example, in the following correspondence theory definition of truth:

x is true:= There is an entity y such that x corresponds to y.

Here the one-place predicate 'x is true' is defined by means of the two-place predicate 'x corresponds to y'. Since the one-place predicate is defined by a two-place predicate it is a *relational* predicate.

Thus, for example, according to Frege the truth of an idea depends on something distinct from the idea, and so that truth is relativized in this sense. For it is now *possible*, at least in principle, of one and the same idea of a church to be true in that it corresponds to the Cologne Cathedral, yet at the same time not true in that it fails to correspond to St Peter's Cathedral in Rome. Thus, the truth of the idea (or its falsity) does not pertain to the idea *per se* but only to the idea relative to one church or the other. Truth in this sense is relative.

If Tarski holds a correspondence theory of truth as his informal remarks seem to suggest, then his is a relational concept of truth. This, at any rate, is how he has generally been understood in philosophical scholarship.[1]

(2)   A Manifold of Truth Concepts

The concept of truth is relativized in a different sense from the one outlined above when we replace the one truth predicate 'X is (absolutely) true' with a *family of one-place truth predicates* 'X is true in S_1', 'X is true in S_2' and so on. Here S_1, S_2, ... may be different natural languages or different formal languages belonging to a Tarskian hierarchy of object- and meta-languages.

Replacement of the *one* (absolute) truth concept by such a manifold of other truth concepts is precisely what we normally mean when we speak of a relativization of the concept of truth. For, as a consequence of such replacement, it becomes *possible* – at least in principle – for one and the same sentence to be true in S_1 yet false in S_2. Truth in this sense is relative as well.

In his formal definition of truth Tarski replaces Frege's *one* absolute concept of truth with a *multitude* of truth concepts, thus relativizing the concept of truth in the sense described above – or at least, this is what we would have to conclude from Frege's perspective.

If we assume a multitude of truth concepts 'x is true in S_1', 'x is true in S_2', ... that are assigned to the various languages S_1, S_2, ..., then we can also define a relational truth concept as follows:

x is true:= There is a language S such that x is true in S.

In this case the newly defined truth concept 'x is true' is relational, but the given truth concepts 'x is true in S_1', 'x is true in S_2', ... do not have to be relational truth concepts at all.

If, conversely, we assume a relational truth concept

x is true:= There is a y so that x stands in relation R to y.

as well as a multitude of objects O_1, O_2, ... to qualify as the second relatum, then we can also define a multitude of truth concepts as follows:

x is true$_1$:= x stands in relation R to O_1.

x is true$_2$:= x stands in relation R to O_2.

...

Thus, under certain circumstances we can switch from the one kind of relativization to the other – though we do not *have* to do so. Having a multitude of truth concepts does not force us to accept a relational truth concept, and vice versa.

(3) Relativistic truth

Often the two previously mentioned kinds of relativization are rejected on the grounds that they might lead to the position of *relativism*, which dispenses with generally authoritative judgments altogether. This position claims as a *fact* (not only as a possibility) that the same sentence is true in S_1 yet false in S_2, or that the same idea is true relative to this object yet false relative to that object. Basically, it would allow us to make judgments about truth or falsity just as we please.

Yet such relativistic consequences cannot be derived from either of the previously discussed concepts of relativization. When setting up a Tarski-style language hierarchy we can ensure that no sentence has different truth-values in two languages within the hierarchy. If a sentence 'p' occurs in two natural languages S_1 and S_2 with different truth-values, then we can ask, first of all, in which of these languages we should interpret it. Presuming that S_1 is this language, then the sentence – as interpreted in S_1 – has only one truth-value. Thus, relativism can be avoided in this case as well. Analogously we can ask – as Frege does – about the church example given in (1) above, what church the picture is supposed to depict; and if this church is Cologne Cathedral then it would be absurd to say that it is a false picture of St Peter's Basilica in Rome.

Neither of the first two kinds of relativization necessarily forces us to accept a *relativistic* concept of truth. Nonetheless, from Frege's perspective I should like to speak of a *relativization* of truth – both in the case of a relational concept of truth and in the case of a multitude of truth concepts; for both are unable to accommodate the *one* absolute and entirely *independent* truth, which is the crucial point of concern to Frege.

If in the following I mention 'Tarski's relative concept of truth', I mean not only the kind of relativization that is caused by a multitude of truth concepts, but also the widespread reception of Tarski's theory of truth as a correspondence theory. In other words, I call Tarski's concept of truth relative in the sense of (1) or (2), but not in the sense of (3).

2.4   Regions of truth

2.4.0   After profiling the word 'true' to 'point the way for logic' in the first paragraph, Frege then in the second paragraph raises the question of how he wants to 'use "true" in this connection'. Frege carefully distinguishes the kind of truth that is at the heart of logic from other, logically irrelevant meanings of the word 'true' such as 'genuine', 'veracious', and so on. What he has in mind is 'that sort of truth which it is the aim of science to discern [in short: *scientific truth*]' (Frege, 1918a, p. 59). In the following, whenever I use the expression 'scientific truth' I implicitly refer to this quotation.

At the beginning of the *third* paragraph we finally encounter the *question of truth bearers* that shall remain the focus of our subsequent discussion:

> Grammatically, the word 'true' appears as an adjective. Hence, the desire arises to delimit more closely the region within which truth could be predicated, the region in which the question 'Is it true?' could be in principle applicable. (i)[2]

The answer to this initial question of the *third* paragraph is given to us no sooner than at the end of the *fourth* paragraph:

> And hence the only thing for which the question 'Is it true?' can be in principle applicable is the sense of sentences. (ix)

2.4.1   The close connection between the *question* of 'the region in which the question "Is it true?" could be in principle applicable' ['das Gebiet [...], wo überhaupt Wahrheit in Frage kommen könne'] and its *answer* that the sense of sentences is 'the only thing for which the question "Is it true?" can be in principle applicable' ['dasjenige, bei dem das Wahrsein überhaupt in Frage kommen kann'] is indicated in their very wordings: the answer almost literally refers back to the question. Question and answer together fuse the third and fourth paragraphs into one argumentative unit. The objections that Frege raises in the third paragraph against the correspondence theory should therefore be read in the context of this question: how do these objections contribute to answering the question raised above regarding the nature of the truth bearers?

Previous interpretations have disregarded both the question and the answer, thus completely ignoring the significance of the fourth paragraph for Frege's overall argumentation. According to such readings of the text, the circle argument at the end of the third paragraph delivers Frege's major conclusion. There Frege argues, though rather by way of an excursion, for the fundamental indefinability of truth – which certainly comes across as a spectacular thesis after Tarski's definition of truth. But if we place this thesis at the center of Frege's argumentation, then we considerably distort the latter's original structure.

2.4.2   Now, at first glance, truth is predicated of extremely diverse types of things, namely 'pictures, ideas, sentences and thoughts' (i). Although we cannot, according to Frege, actually define the sense of the word 'true', we can nevertheless tell that these diverse types of things must be called true in different respective senses. For pictures, ideas, sentences, and thoughts are each so different from the others that they cannot all be true in the same sense of the word: '[...] alterations in sense have taken place' (i).

In view of the diversity of putative truth bearers, the above question must be reformulated more precisely as: of what is truth *really* predicated? Frege answers this question by clarifying the logical connections between the various truth concepts associated with the various 'regions of truth', with the result that the truth of *senses* (of sentences) emerges as the basic or 'real' concept of truth: the truth of an idea/a picture is reduced to that of a sentence, and the truth of a sentence is reduced to the truth of its sense, that is of the thought expressed in it.[3]

Though Frege is unable to provide a definition of the scientific concept of truth (see Chapter 8), he nonetheless attempts to achieve a certain degree of unambiguousness by delimiting the region of truth – the region, in other words, of objects of which the scientific concept of truth is really predicated.

2.5   Two notions of truth

Frege starts out with four potential candidates for this region of basic objects of truth, namely pictures, ideas, sentences, and thoughts. Subsequently, however, he combines the first two and the last two, respectively, thereby reducing the distinction to essentially *two* different concepts of truth: On one hand, there are pictures and ideas, whose truth we shall hereafter call 'I-truth', and on the other hand, there are sentences and their senses (thoughts), whose applicable truth-concept we shall call 'S-truth'.

Frege describes the connection between an idea and a picture in 'Logik' as follows:

> By an idea we understand a picture that is called up by the imagination: unlike a perception it does not consist of present impressions, but of the reactivated traces of past impressions or actions. Like any other picture, an idea is not true in itself, but only in relation to something to which it is meant to correspond. (Frege, 'Logik', p. 142)

According to Frege ideas are but a kind of picture and their truth consists in correspondence to something else.

With regard to S-truth, Frege distinguishes further between a logically primary and a logically secondary concept by reducing sentence-truth to sense-truth: 'And when we call a sentence true we really mean that its sense is true' (ix). As this subdivision within the concept of S-truth is unproblematic for Frege we shall not discuss it further here; instead, in what follows we shall generally speak of *the* concept of S-truth, without noting any further differentiation.[4] While Frege does not offer a single definition of S-truth in the third and fourth paragraphs, he does provide at least three definitions of I-truth in the context of his first three objections – which serve to refute each one of these same definitions. It is important to note, however, that the definitions do not fail in themselves – that is, not as definitions of I-truth as such; rather, what Frege shows is that I-truth as defined in any of these ways fails to satisfy certain requirements of the *scientific* concept of truth. In other words, they fail as definitions of scientific truth.

Because I-truth turns out to be unsuitable as a conception of scientific truth, only S-truth remains a potential candidate for this role. Having thus identified S-truth as the real and logically primary concept of truth, Frege is now also able to explain the alterations in sense: 'So what is improperly called the truth of pictures and ideas is reduced to the truth of sentences' (viii). In short: I-truth is reduced to S-truth.

Though at the very end of his argument Frege calls I-truth an *improper* kind of truth, the distinction between the *two* truth concepts is essential to the entire point of the third and fourth paragraphs – for that point is precisely to decide *which* of the two kinds of truth qualifies as scientific truth. In other words, the issue is about the proper bearers of scientific truth.

2.6 The truth of pictures and ideas in terms of correspondence

2.6.0 Ideas and pictures are, first of all, things in the world just like stones and leaves. Then why are we sometimes tempted to call a picture true but never a stone? The reason is that a picture comes with an intention: 'A picture is meant to represent something' (iii), and if it succeeds in fulfilling this intention then we call it 'true'. Like pictures, ideas too involve intentions; they are supposed to 'correspond to something' (iii), and are called 'true' if this correspondence obtains. By contrast, we generally do not associate intentions with stones or leaves, and this is why we do not consider them as true or false.

In the third paragraph Frege defines exclusively I-truth – that is, the truth of pictures and ideas – in terms of correspondence. In the following, I shall quote in order the only sentences from the third paragraph that contain, in addition to the word 'correspondence' (or 'correspond'), one or the other of the four words 'picture', 'idea', 'sentence', and 'thought':

> [(1)] Neither is an *idea* called true in itself, but only with respect to an intention that it should *correspond* to something. It might be supposed from this that truth consists in a *correspondence* of a *picture* to what it depicts. (iii)

The last sentence contains the assumption refuted by Frege's critique of the correspondence theory in the third paragraph. The following partial sentence belongs to Frege's second argument:

> [(2)] [...] when they define truth as the *correspondence* of an *idea* with something real. (v)

And the following partial sentence belongs into the context of the third argument:

> [(3)] [...] whether it were true that, for example, an *idea* and something real *correspond* in the specified respect. (vi)

The following sentences from the beginning of the fourth paragraph match these quotations:

> [(4)] [...] we want to say that that *picture corresponds* in some way to this object. 'My *idea corresponds* to Cologne Cathedral' [...]. (viii)

Note that on one hand, Frege appears in (1) and (4) to regard the words 'idea' and 'picture' as interchangeable; on the other hand, nowhere in the third paragraph is the concept of correspondence applied to sentences or thoughts. This is why we should also read the phrase 'that, for example, an *idea* [...]' in (3) in such a way that the word 'idea' could there be replaced by 'picture' without changing the point that Frege is trying to make, though not in such a way that the word could be similarly replaced by 'sentence' or 'thought'. The interchangeability of 'picture' and 'idea' on one hand, and their strict differentiation from sentences and thoughts on the other, is also supported by the following (previously quoted) sentence from the fourth paragraph: 'So what is improperly called the truth of pictures and ideas is reduced to the truth of sentences' (viii).

   Thus, in the third paragraph Frege defines exclusively *I-truth* in terms of correspondence. Accordingly, his first three arguments address the correspondence theory solely as a theory of the truth of *ideas and pictures*. This is not to imply that Frege's objections could not in principle be applied to a correspondence theory for sentences as well (see section 13.2 and Chapter 14). However, my point is that these objections should be first studied within the original argumentative context in which they were made in order to be properly understood.

2.6.1   The only sentence in either the third or fourth paragraphs that contains both the word 'correspondence' and the word 'sense', is at the very end of the fourth paragraph; it reads,

> In any case, truth [of the sense of a sentence] does *not* consist in correspondence of the sense with something else, for otherwise the question of being true would get reiterated to infinity. (x)

Here, the concept of correspondence is applied, albeit negatively ('not'), to a sense and not to an idea. However, this passage contains Frege's regress argument, which is very different from his first three arguments against the correspondence theory in the third paragraph. In particular, Frege's regress argument presupposes that one of the two concepts of truth can be reduced to the other, which is argued by Frege no earlier than at the beginning of the fourth paragraph (see Chapter 11). This is why the above quote from the end of the fourth paragraph is consistent with my thesis that in the first three arguments in the *third* paragraph Frege defines only *I-truth* in terms of correspondence.

3
Frege's Text and Its Argumentative Structure

3.1 The text

3.1.0 In the following chapters I present a close reading of the actual text of the third and fourth paragraphs of Frege's 'Der Gedanke'. Toward this end I begin by quoting the text of both paragraphs in full and making the immanent argumentative structure of the text explicit. I have structured the text into sub-paragraphs (i) – (x) and added headings as well as insertions.

3.1.1 The initial question: of what can scientific truth be predicated – of ideas (pictures) or of thoughts (sentences)?

[(i)] Grammatically, the word 'true' appears as an adjective. Hence, the desire arises

[(α) Initial question:] to delimit more closely the region within which truth could be predicated, the region in which the question 'Is it true?' could be in principle applicable. We find truth predicated of pictures, ideas, sentences, and thoughts. It is striking that visible and audible things occur here along with things which cannot be perceived with the senses. This suggests that alterations in sense have taken place.

The truth of pictures and ideas (I-truth) as correspondence.

[(ii)] Indeed they have! For is a picture, as a mere visible and tangible thing, really true? And a stone, a leaf is not true?

[(iii)] Obviously, we would not call a picture true unless there were an intention involved. A picture is meant to represent something.

17

Neither is an idea called true in itself, but only with respect to an intention that it should correspond to something.

[(β) Assumption:] It might be supposed from this that truth consists in a correspondence of a picture to what it depicts.

The first objection: scientific truth is absolute.

[(iv)] Now a correspondence is a relation. But this goes against the use of the word 'true', which is not a relation word, does not contain any indication of anything else to which something is to correspond. If I do not know that a picture is meant to represent Cologne Cathedral then I do not know what to compare the picture with in order to decide on its truth.

The second objection: scientific truth is perfect (it does not admit of gradations).

[(v)] A correspondence, moreover, can only be perfect if the corresponding things coincide and so just are not different things. It is supposed to be possible to test the genuineness of a banknote by comparing it stereoscopically with a genuine one. But it would be ridiculous to try to compare a gold piece stereoscopically with a twenty-mark note. It would only be possible to compare an idea with a thing if the thing were an idea too. And then, if the first [idea] corresponds perfectly with the second [idea], they coincide. But this is not at all what people intend when they define truth as the correspondence of an idea with something real. For in this case it is essential precisely that the real thing be distinct from the idea. But then there can be no perfect correspondence, no perfect truth. So nothing at all would be true; for what is only half true is untrue. Truth does not admit of more or less.

The third objection: scientific truth is independent.

[(vi)] Or does it? Could we not maintain that there is truth when there is correspondence in a certain respect? But which respect? And what would we then have to do so as to decide whether something were true? We should have to inquire whether it were true that, for example, an idea and something real correspond in the specified respect. And with that we should be confronted again by a question of the same kind, and the game could start all over.

[(ββ) Final conclusion of the refutations:] So this attempt [extending through (iv) – (vi)] to explain truth as correspondence [of ideas or pictures with something in reality] breaks down.

The fourth objection (an excursion): applying an analytic definition of S-truth leads to a vicious circle.

[(vii)] [Conclusion of the circle argument:] But likewise, any other attempt to define truth also breaks down.

For in a definition certain characteristics [note the plural!] would have to be specified. And in application to any particular case it would always depend on whether it were true that the characteristics were present. So we should be going round in a circle.

[Final conclusion of all four objections:] So it is likely that the content of the word 'true' is sui generis and indefinable. [End of the excursion.]

The answer to the initial question: truth is predicated of the sense of sentences.[1]

[(viii)] When we predicate truth of a picture we do not really mean to predicate a property which would belong to this picture altogether independently of other things. Rather, we always have in mind some totally different object and we want to say that that picture corresponds in some way to this object. 'My idea corresponds to Cologne Cathedral' is a sentence, and now it is a matter of the truth of this sentence. So what is improperly called the truth of pictures and ideas is reduced to the truth of sentences.

[(ix)] What is it that we call a sentence? A series of sounds, but only if it has a sense, which is not to say that any series of sounds that has a sense is a sentence. And when we call a sentence true we really mean that its sense is true.

[(αα) The answer to the initial question] And hence the only thing for which the question 'Is it true?' can be in principle applicable is the sense of sentences [in short, the thought].

The fifth objection: Frege's regress argument.

[(x)] Now is the sense of the sentence an idea? In any case, truth does not consist in correspondence of the sense with something else, for otherwise the question of whether something is true would get reiterated to infinity.

3.2   Discussion of the structure: demarcating the arguments

3.2.0   The argumentative structure proposed in the previous section is based first and foremost on the central *question* in the third and fourth paragraphs:

> [(α)] [How] to delimit more closely the region within which truth could be predicated, *the region in which the question 'Is it true?' could be in principle applicable.* (i)

The *answer* to this question is given only at the end of the *fourth* paragraph:

> [(αα)] And hence *the only thing for which the question 'Is it true?' can be in principle applicable* is the sense of sentences.

As pointed out earlier, the close connection between question and answer is clearly indicated by the similarity in wording (see the italicized phrases in the quotations above). Question and answer together fuse the third and fourth paragraphs into one argumentative unit. Without the fourth paragraph, which has not been considered at all in previous interpretations, we could not properly understand the third one, either.

3.2.1   Frege's refutation of the correspondence theory in the third paragraph is limited by the assumption that I-truth may be scientific truth and by the conclusion of the related indirect proof. The *assumption* is:

> [(β)] It might be supposed from this that *truth consists in a correspondence* of a picture to what it depicts.

Frege then discusses *three* definitions of I-truth, which supplement one another so as to form, jointly, *one* adequate definition of I-truth. But none of these definitions, nor the combination of them, turns out to be adequate as a definition of scientific truth. For according to Frege, scientific truth is *absolute, perfect,* and *independent,* while I-truth – as Frege shows in three objections – is none of these things. With these three objections, assumption (β) is refuted as well. Frege formulates the final conclusion by reference to his original wording of assumption (β) (see the italicized parts):

[(ββ)] So this attempt [extending through (iv) – (vi)] to explain *truth as correspondence* breaks down.

According to my reading (indicated in bracketed remarks), the expression 'this attempt' refers to *all three* definitions together (see below), which can be combined into *one* comprehensive definition of I-truth, as an attempt to explain 'truth as correspondence'. Dummett and other scholars, however, understand the expression 'this attempt to explain' [Geach: 'the attempted explanation'] as referring only to the third definition in (vi).

3.2.2   There is an excursion in the third paragraph directly following Frege's refutation of the correspondence theory of I-truth; in this excursion Frege argues for the indefinability of S-truth.

   While the third paragraph shows that I-truth cannot be scientific truth – and hence ideas cannot serve as truth bearers in science – the fourth paragraph contains a reduction of I-truth to S-truth. Only now does Frege answer the initial question (α) in terms of (αα). The argumentation is rounded off with a regress argument. This regress argument is completely different from Dummett's regress argument (cf. Chapters 10 and 12).

3.2.3   My way of demarcating the four arguments (iv) – (vii) differs from that of other interpreters and hence demands further justification. First of all, I look at the beginnings of the arguments and subsequently at their respective endings. Though we might be tempted to assume that the one simply follows from the other, the matter is not that simple in this case.

   The first argument (see (iv)) starts immediately after the sentence in which the indirect assumption is formulated: (β) 'It might be supposed from this that truth consists in a correspondence of a picture to what it depicts' (iii). Frege uses some obvious linguistic markers to signal the beginnings of the subsequent arguments. The second argument (see (v)) starts with 'auch' ('moreover') and then introduces the concept of perfect truth, which is the topic of only this argument, thereby clearly demarcating that argument from the first and the third ones: 'A correspondence, *moreover*, can only be *perfect* if [...].' The third argument (see (vi)) starts with the question: 'Or does it?' ('Oder doch?'), and the fourth argument (see (vii)) starts with 'any *other* attempt' ('jeder *andere* Versuch'): 'But likewise, any *other* attempt to define truth also breaks down.'

Hence, we have established that Frege fairly clearly distinguishes the four objections (iv) – (vii) from one another by signaling the beginning of each new argument using obvious verbal markers.

3.2.4   The concluding passages of the four arguments are not quite as obvious as the beginnings, since in addition to the absurd consequences with which all of the arguments end, they sometimes also contain the results of the arguments, and it is not always clear at first sight what these results refer to. Let me first quote the absurd consequences that appear at the end of each of the *reductio ad absurdum* arguments:

> [...] I do not know what to compare the picture with in order to decide on its truth. (iv)

> So nothing at all would be true; for what is only half true is untrue. Truth does not admit of more or less. (v)

> [...] the game could start all over. (vi)

> So we should be going round in a circle. (vii)

With these absurdities, the respective argumentative goals of the four arguments are achieved. Though there are *four* arguments, there are only *three* sentences in which the respective results are formulated:

> [(A)] So *this* attempt to explain truth as correspondence breaks down. [Last sentence in (vi)]

> [(B)] But likewise, any *other* attempt to define truth also breaks down. [First sentence in (vii)]

> [(C)] So it is likely that the content of the word 'true' is [...] indefinable. [Last sentence in (vii)]

(B) anticipates the result of the fourth argument. This is followed by the justification starting with 'for' ('denn'). Thus it seems clear to me where (B) belongs.

Which argument exactly is being referred to in (A), however, is in dispute. If we quote the third and fourth arguments in isolation from the first and second arguments, as is usually done, then we are compelled to refer the expression 'this attempt' to the third argument, because in that case there is no other, alternative argument (in the quotation) to which we could refer the expression. As a result of this way of quoting we will understand the sentences

(A') *So* this attempt breaks down,

(B') But *likewise*, any other attempt breaks down.

such that the respective 'So [...] breaks down' or 'But likewise [...] breaks down' in these two sentences means that it is *the same* argument that makes both attempts break down. It is in this spirit that Dummett paraphrases sentence (B) as 'The same reasoning shows [...]' (see Chapter 10). But if Frege's third and fourth arguments essentially are the same argument, it is certainly justified to quote these two arguments as one argumentative unit, while leaving out the first two arguments. Thus this way of quoting basically sustains itself.[2]

3.2.5   Once we look at the complete text, however, it no longer appears plausible to link the expression 'this attempt' exclusively to the third argument, since doing so would rule out the possibility that the first three arguments end up in the same way. Instead, the first and second arguments would end up with the demonstration of an absurd consequence, while the third one would, *in addition* to its demonstration of an absurdity ('the game could start all over') *also* provide a formulation of the result ('So this attempt [...] breaks down'). But the attempts under scrutiny in the first and second arguments also break down, just like this one; why would the breaking down of the third attempt deserve more attention than the others?

Just like the third argument, the first and second arguments refute attempts at explaining truth as a form of correspondence; likewise, all three arguments provide critiques of different variants of the correspondence theory. It is therefore plausible to see the expression 'this attempt' as referring jointly to the first three arguments – as referring, that is, to the entire attempt comprising (iv) – (vi) 'to explain truth as correspondence'. This is also consistent with the above demarcation of Frege's discussion of the correspondence theory by the following two sentences:

[(β)] It might be supposed from this that *truth consists in a correspondence* of a picture to what it depicts.

[($\beta\beta$) = (A)] So this attempt [extending through (iv) – (vi)] to explain *truth as correspondence* breaks down.

According to this interpretation of 'this attempt', the first three arguments have the same kind of ending, that is, they all end with the derivation of an absurd consequence, only after which the result of all three variants combined is presented ('So this attempt [...] breaks down').

If this is correct, the sentence 'So this attempt [...] breaks down' in (A) cannot mean 'This attempt breaks down due to *this* argument', for altogether *three distinct* arguments were presented against the three variants of the correspondence theory. Nor could we paraphrase the sentence 'But likewise, any other attempt breaks down' in (C) as, 'But likewise any other attempt breaks down due to the *same* argument' (see Dummett: 'The same reasoning shows...'), for there are three different arguments preceding this – none of which constitutes a circle argument, as the fourth one does.

3.2.6 If the expression 'So' ('Hiernach') in the sentence (C) 'So it is likely that the content of the word "true" is sui generis and indefinable' referred exclusively to the fourth argument, the circle argument, then the result of the fourth argument would be formulated twice, since the circle argument already starts out with an anticipation of the result: (B) 'But likewise, any other attempt to define truth also breaks down.' The anticipated and the retrospectively formulated results of the circle argument, however, are not quite consistent with each other. According to (C) the circle argument would rule out *all* definitions of truth, while according to (B) they would rule out only *all others* (we would have to ask: other than what?).

In (A) Frege formulates a summary result of the first three arguments; analogously the reader expects a summary result of the first four arguments at the end of the third paragraph. If the sentence (C) referred exclusively to the circle argument, then a summary result of the first four arguments would be entirely missing; instead, we would have two incompatible formulations of the results of the fourth argument.

This makes it more plausible to assume that sentence (C) does not refer exclusively to the circle objection but to all four arguments. Thus we obtain the following allocation of results to arguments:

Sentence (A) summarizes the results of the first three arguments.

Sentence (B) anticipates the result of the fourth argument.

Sentence (C) summarizes the results of all four arguments.

3.3   Three definitions of I-truth in terms of correspondence

As we can see from Frege's definitions of I-truth, he starts out from a three-place concept of correspondence:

x corresponds to y with respect to H.

The text offers the following proposals for a definition of I-truth in terms of correspondence. The first definition is given in sentence (β) of passage (iii) and explains I-truth as follows:[3]

(D₁) Idea x is true:= Idea x *corresponds to* the depicted object.

This definition becomes the basis for the two subsequent definitions. In passage (v) Frege emphasizes the necessary *distinctness* of idea and depicted object. Thus the second definition goes:

(D₂) Idea x is true:= Idea x corresponds to the depicted object, which is *distinct from* x.

Finally, the first definition is supplemented by the *respect* in which the correspondence is to hold (see passage (vi)). This gives rise to the third definition:

(D₃) Idea x is true:= Idea x corresponds to the depicted object *in a certain respect H**.

All three definitions together can be combined to *one* comprehensive definition of I-truth:

(D₁₋₃) Idea x is true:= Idea x corresponds to the depicted object, which is distinct from x, in a certain respect H*.

According to Frege, definitions (D₁) – (D₃) and certainly also definition (D₁₋₃) are adequate with regard to I-truth. On several occasions, he writes – without criticism – that a picture intended to depict Cologne Cathedral corresponds to the latter or, alternatively, that my idea of Cologne Cathedral corresponds to that building: '"My idea corresponds to Cologne Cathedral" is a sentence, and now it is a matter of the truth of this sentence' (viii). Nonetheless, for Frege I-truth is not suitable as a candidate for *scientific* truth. In the following reconstruction of Frege's three objections against the correspondence theory precisely this question shall take center stage: Can I-truth, thus defined in terms of correspondence, satisfy the requirements that we associate with our scientific concept of truth?

Another way to put this question makes its relevance to psychologism even more explicit: can pictures or ideas, whose truth can only be understood in terms of their correspondence with something else, be bearers of scientific truth? For Frege, a declared opponent of psychologism, the answer can only be: 'No!'

4
The First Argument: Scientific Truth Is Absolute

4.1 The text

(iv) Now a correspondence is a relation. But this goes against the use of the word 'true', which is not a relation word, does not contain any indication of anything else to which something is to correspond. If I do not know that a picture is meant to represent Cologne Cathedral then I do not know what to compare the picture with in order to decide on its truth.

4.2 The absoluteness of scientific truth

4.2.0   Frege introduces the notion of I-truth in the following passage:

Obviously, we would not call a picture true unless there were an intention involved. A picture is meant to represent something. Neither is an idea called true in itself, but only with respect to an intention that it should correspond to something. (iii)

Here I-truth is clearly distinguished from truth in itself; an idea is 'not true in itself'. Frege makes the same distinction at the beginning of the fourth paragraph:

When we predicate truth of a picture we do not really mean to predicate a property which would belong to this picture altogether independently of other things. (viii)

Ideas and pictures are *not* true in themselves; their truth does *not* belong to them 'independently of other things'. By characterizing I-truth

26

negatively in this way, Frege at the same time gives us a clue as to the positive concept of truth he is looking for. What is true in the sense of science, is 'true *in itself*', that is, true *absolutely* or '*independently* of other things'.

4.2.1 The passage quoted above as (iii) contains the following definitions of I-truth in which Frege defines the one-place predicate 'x is true' by means of the two-place predicate 'x corresponds to y':

> (D_0) Idea x is true:= Idea x corresponds to the object to which it is intended to correspond (see the sentence just before (β)).

> (D_1) Idea x is true:= Idea x corresponds to the depicted object (see (β)).

Definition (D_0) is unacceptable for science because, first of all, it presents truth as being dependent on intentions. But Frege's objection here is even more general and applies to both definitions:[1]

> Now a correspondence is a relation. But this goes against the use [as common in science] of the word 'true', which is *not a relation word*, *does not contain any indication of anything else* to which something is to correspond. If [with respect to the truth of a picture] I do not know that a picture is meant to represent Cologne Cathedral then I do not know what to compare the picture with in order to decide on its truth. [Thus, the truth of a picture requires an 'indication of something else', and in *this* use of the term, 'truth' is a relation word.] (iv)

If the picture of Cologne Cathedral is true in the sense of (D_1), then it corresponds to the Cathedral or, in other words, it is *true of* the Cathedral – in any case, it is not true *in itself.* Truth in the sense of (D_1) is always *truth-of* or *truth in relation to* something, just as a wife by definition is wife *of* or *in relation to* a certain man and a father is by definition father *of* or *in relation to* a child. In science, however, we use the term 'true' in a different sense from that expressed in (D_1). If it is true that $1 + 1 = 2$ then it is true *in itself,* not *of* something or *in relation to* something (whether that be the number 1 or the number 13). Likewise, if it is true that Cologne Cathedral has two steeples then this is true in itself, not *of* something (whether that be Cologne Cathedral or Cologne Central Station). This is, in my view, what Frege means when he says that the term 'true' is (in its scientific use) 'not a relation word'.

4.2.2   The expression 'not a relation word' is contained in the following sentence:

> But this goes against the use of the word 'true', which is not a relation word, [that is,] does not contain any indication of anything else to which something is to correspond. (iv)

Frege here explains what he means by a 'relation word' in this context:

> 'True' is *not* a relation word iff it 'does *not* contain any indication of anything else to which something is to correspond.'

Hence,

> 'true' *is* a relation word iff it *does* contain an indication of something else to which something is to correspond.

According to this quoted passage a *relation word* in Frege's sense contains an indication of something else: it denotes a *relational property*, as we would call it today (see section 13.2). The one-place predicate 'x is true' contains an indication of something else, if it is understood in the sense of the following definitions:

> (D$_0$) Idea x is true:= Idea x corresponds to the object to which
> it is intended to correspond.

> (D$_1$) Idea x is true:= Idea x corresponds to the depicted object.

According to Frege 'true' (in the scientific sense) is not a relation word, because 'it does *not* contain any indication of anything else to which something is to correspond' and hence it does not denote a relational property. One-place predicates, which do not denote a relational (or relative) property, denote an *absolute* property. Such predicates may be called 'absolute predicates'.

   Though Frege does not literally use the German equivalent of the word 'absolute' ('absolut') in his original text, it is clear that he means what is commonly understood as absolute, as opposed to relative truth. On this issue, I side with Stepanians, who considers truth in Frege as in a specific sense absolute (Stepanians, 2001, p. 155). However, Stepanians does not apply this insight to his reading of Frege's first objection (see section 14.2). We could ask why Frege does not actually use the word 'absolute' in this context, if this is what he really means. I suspect that for Frege the word 'absolute' was unpalatable due to its association with Hegel's and Fichte's Philosophy of the Absolute.

4.2.3   What is decisive is the *use* of the word 'true' in the context of science. Both Frege and the psychologists use the predicate 'x is true' as a one-place predicate; but while the psychologists can define their meaning of 'x is true' by means of the two-place predicate 'x corresponds to y', Frege considers such a definition as 'improper' (viii) with respect to scientific truth. Indeed, the term 'Gebrauchsweise' ('use' in English) occurs twice in the second paragraph in order to exclude certain meanings of 'true' right at the outset of Frege's investigation.

> [...] first I shall attempt to outline roughly how I want to use 'true' in this connection, so as to exclude irrelevant *uses* of the word. 'True' is not to be used here in the sense of 'genuine' or 'veracious'; [...] This *use* too [as in 'a true friend'] lies off the path followed here. What I have in mind is that sort of truth which it is the aim of science to discern. (Frege, 1918a, p. 59; my italics)

In the third and fourth paragraphs Frege asks about the scientific concept of truth and its *use* ('Gebrauchsweise'): do we predicate truth in the scientific sense to pictures and ideas or rather to sentences and their senses?

4.2.4   Frege's first argument can be summarized as follows:

> (P1) I-truth is correspondence.
>
> (P2) If I-truth is correspondence then it is relative (relational) truth.
>
> (P3) Scientific truth, however, is absolute (i.e. not relative).
>
> (C) Therefore, I-truth cannot be scientific truth.

4.3   Soames's and Künne's criticisms of Frege's first argument

4.3.0   Soames's reconstruction of Frege's first argument and his critique are as follows:

> The argument to this point is quite simple:
> 1.   Grammatically, 'true' is a predicate applied to individual objects and so ought to stand for a property if it stands for anything at all.

2.   Correspondence is a relation holding between at least two objects.
3.   Thus truth is not correspondence – that is, the word 'true' does not stand for any correspondence relation (since it does not stand for any relation at all).

This conclusion is both correct and general, applying not just to pictures but to other truth bearers, such as propositions, as well. However, it has no force against correspondence theories of truth. Such theories do not say that 'true' means 'corresponds to' or that the property truth is identical with the relation of correspondence. Rather, they say that truth is a relational property – the property of corresponding to something in reality. Relational properties are very familiar – for example, to be a father is to be a father of someone. The property of being a father, which holds of individuals, is not identical with the relation of fatherhood, which holds between pairs of individuals. Nevertheless, the former can be defined as the property of bearing the latter to some individual. According to correspondence theories, a similar relationship holds between truth and correspondence. Such theories may or may not be correct, but they are not undermined by the observations in (1) – (3). (Soames, 1999, p. 24)

Künne arrives at a very similar assessment:

I quote this ['Now a correspondence is a relation. But this goes against the use of the word "true", which is not a relation word [...]'(modified standard translation)] not because I consider it to be a strong objection [against the correspondence view of truth], but because I think that it reveals how the correspondence slogan is to be taken if it is to have any philosophical bite: it must be understood as declaring truth to be a relational property. [...] As an objection, Frege's argument is rather weak. To be sure, unlike 'agrees with' or 'corresponds to', the predicate 'is true' is a one-place predicate, hence it does not signify a *relation*. But the predicate 'x is a spouse' is also a one-place predicate, hence it does not signify a relation either, and yet it is correctly explained as 'There is somebody to whom x is married'. It signifies a *relational property*. [...] Perhaps the predicate 'x is true' can be similarly explained: 'There is something to which x corresponds'. (Künne, 2003, pp. 93–4)

4.3.1   Thus, we see that both Künne and Soames ascribe to Frege the view that for formal reasons, a *one*-place predicate such as 'x is true' could not, in principle, be defined by means of a *two*-place predicate such as 'x corresponds to y', because these two predicates are of *differ-ent arity*.[2] But this means ascribing to Frege a beginner's mistake that no student would make after their completion of an introductory logic course. Was Frege indeed dumber than the average present-day fresh-man? I find myself having great difficulties believing this.

First, Frege's sense of 'relation word' in this context is different from today's sense. Soames and Künne regard the two-place predicate 'x cor-responds to y' as a relation word, but this predicate 'does *not* contain any indication of anything else to which something is to correspond' and hence 'is *not* a relation word' in Frege's sense. The one-place predicate 'x corresponds to the object to which it is supposed to correspond' is a relation word in Frege's sense and has the same arity as the one-place predicate 'x is true'. Though both predicates have the same arity, accord-ing to Frege the former predicate cannot be the definiens in an adequate definition of (scientific) truth. Therefore, contrary to Soames and Künne, Frege's first argument has nothing to do with difference in arity.

Second, it was Frege who introduced the distinction between one-place and two-place predicates in logic and who therefore should have known that the one-place predicate 'x is a father' can be defined in terms of 'There is a y, such that: x is father of y'. As a mathematician, he should have recognized even in his dreams that the one-place predicate 'x is an even number' can be defined by means of the two-place predicate 'x is divisible by y' as follows:

x is an even number:= x is divisible by 2.

Third, as we saw with regard to definitions $(D_0) - (D_3)$, in the text Frege provides *his own* examples of how to smoothly define the one-place predicate 'x is true' by means of the two-place predicate 'x corresponds to y' or even the three-place predicate 'x corresponds to y with respect to H'. The definition

(D_0) Idea x is true:= Idea x corresponds to the object that it is intended to correspond to.

was derived from what Frege points out in the following passage already quoted:

A picture is meant to represent something. Neither is an idea called true in itself, but only with respect to an intention that it should cor-respond to something. (iii)

Likewise, the definition

(D₁) Idea x is true:= Idea x corresponds to the depicted object.

was based on Frege's assumption (β) 'that truth consists in a correspondence of a picture [an idea] to what it depicts' (iii). Obviously, these passages show how Frege *himself* defines a one-place truth predicate by means of a two-place correspondence predicate. Likewise, the definition

(D₃) Idea x is true:= Idea x corresponds to the depicted object *in a certain respect H**.

can be derived from Frege's proposal 'that there is truth when there is correspondence [to something real] in a certain respect' (vi). Finally, though the definition

(D₂) Idea x is true:= Idea x corresponds to the depicted object, which is *distinct from* x (= which does *not* correspond to x *in every respect*).

is not offered in a single sentence of the text, it is implicitly contained in passage (v) in the context of Frege's second critique of the correspondence theory.

4.3.2   The reasons Frege presents against I-truth as a candidate for scientific truth are content-related rather than formal. Of course, you can formally define a one-place predicate such as 'x is true' by means of a two-place (or even three-place) predicate 'x corresponds to y (with respect to H)'. However, just as we will always fall short of adequately defining the concept of a wife in herself – whatever that might be – by means of the two-place predicate 'x is married to y' and only succeed in thus defining the concept of a wife *of*, or in relation to, somebody, we shall never succeed in adequately defining the concept of truth in itself – which Frege holds to be identical with scientific truth – by means of the two-place predicate 'x corresponds to y'.

In the case of the wife, this does not matter, for every wife is obviously the wife of, or in relation to, somebody; there is no wife in herself, no absolute wife. But in the case of truth, this result is important and relevant, given Frege's claim that there are not only I-truths that depend on something in the sense of (D₁) but also scientific truths in themselves or, in other words, *absolute* truths, just as Kant claims that there is something that is good in itself.

4.3.3   When Künne points out: 'how the correspondence slogan is to be taken if it is to have any philosophical bite: it must be understood as declaring truth to be a relational property' (see above), this means in short:

(a) If truth consists in correspondence then it is a relational property.

The contraposition to this is:

(b) If truth is not a relational (but an absolute) property, then it does not consist in correspondence.

Obviously Künne recognizes (a) and therefore, presumably, (b) as well. With this he comes very close to unwrapping the whole story of Frege's first objection to the correspondence theory. But instead of going all the way and granting Frege his insistence on an absolute concept of truth, he prefers – just like Soames – to attribute to Frege a beginner's mistake that is absurd both in the light of the text itself and in the light of Frege's accomplishments as a logician. It thus appears that in Tarski's shadow the absoluteness of truth is no longer conceivable – not even as a possibility.

5
The Second Argument: Scientific Truth Is Perfect

5.1 The text

(v) A correspondence, moreover, can only be perfect if the corresponding things coincide and so just are not different things. It
is supposed to be possible to test the genuineness of a banknote
by comparing it stereoscopically with a genuine one. But it would
be ridiculous to try to compare a gold piece stereoscopically with
a twenty-mark note. It would only be possible to compare an idea
with a thing if the thing were an idea too. And then, if the first corresponds perfectly with the second, they coincide. But this is not at
all what people intend when they define truth as the correspondence of an idea with something real. For in this case it is essential
precisely that the real thing be distinct from the idea. But then there
can be no perfect correspondence, no perfect truth. So nothing at
all would be true; for what is only half true is untrue. Truth does not
admit of more or less.

5.2 The perfection of scientific truth

5.2.0   I-truth consists not merely in correspondence, but in the correspondence between *distinct* relata. According to Frege, for I-truth it is
'essential precisely that the real thing be distinct from the idea' (v). Idea
and depicted object, though related through correspondence, must not
be numerically identical. Thus we are given the following amendment
to Frege's first definition of I-truth:

(D$_2$) Idea x is true:= Idea x corresponds to the depicted object, which
is *distinct from* x.

Obviously, since idea and depicted object are distinct they cannot correspond in *every* respect. Thus, their correspondence is imperfect; and because I-truth consists in correspondence, it is necessarily an imperfect kind of truth. Scientific truth, by contrast, is *perfect*, as Frege now argues. It follows that I-truth cannot be scientific truth or else 'nothing at all would be true' (v).

5.2.1 If, on the other hand, correspondence were perfect, then the relata would have to be identical; thus, the perfect truth of an idea would amount to nothing but the identity of the idea with itself. But would such perfect truth still qualify as relational truth as *opposed to* absolute truth? Probably not, for even what is absolutely true – whatever this might be – is always identical to itself. Thus, perfect correspondence would be a form of relational truth that is consistent with absolute truth.

However, we may not want to call such truth 'relational' in any meaningful sense of the word. For relational truth in any genuine sense of the word implies a dependence on something other than the truth bearer itself. Even in the first argument Frege presupposes, implicitly, a distinctness between the corresponding relata when he points out that the term 'true', as it is commonly used in science, 'is not a relation word, does not contain any indication of anything else to which something is to correspond' (iv). Thus, already the *absoluteness* of scientific truth is incompatible with the distinctness of the relata.

The second argument differs from the first in that here, the distinctness of the relata is *explicitly* discussed as an integral part of the notion of truth as correspondence, clearing the stage for the property of *perfection* (not admitting of 'more or less' (v)), as a further characteristic of scientific truth in opposition to I-truth.

In addition, the second argument for the first time addresses the *respects* in which the correspondence is to hold: perfect correspondence is correspondence in *every* respect. But I-truth cannot consist in such perfect correspondence. Accordingly, the next question for Frege is whether or not I-truth might, after all, be defined in terms of correspondence in a *certain* respect. And this question is addressed in his third objection.

5.2.2 The second argument can be summarized as follows:

(P1) I-truth is correspondence between distinct relata.

(P2) If I-truth is correspondence between distinct relata then it is an imperfect kind of truth.

(P3) Scientific truth, however, is perfect.

(C) Therefore, I-truth cannot be scientific truth.

5.3   Soames's reconstruction of the second argument

Immediately after the passage quoted in section 4.3, Soames writes:

> After making these observations [see the Soames quote in section 4.3], Frege gives a general argument directed, in the first instance, against all attempts to define truth in terms of correspondence and, by implication, against all definitions of truth whatsoever. He begins by noting that the truth of p cannot involve complete correspondence between p and any fact since complete correspondence is identity and, according to the correspondence theory, a truth is supposed to be distinct from the fact it corresponds to. He goes on: [...] (Soames, 1999, p. 24)

The quoted passage continues with a quotation of Frege's third and fourth arguments.

Soames addresses Frege's second argument only in the second sentence of the above passage but does not even mention it anywhere else in the unquoted portions of his text. Soames apparently reads the argument as a 'general argument directed [...] against all definitions of truth whatsoever': 'He [Frege] *begins* [...]. He *goes on* [...]' (see above, my emphasis). However, Soames fails to specify either in the lines just quoted or in the lines that follow them just what the second argument's contribution to the alleged 'general argument' is supposed to be. According to Soames, Frege begins with an apparently undisputed assumption ('by noting') 'that the truth of [proposition] p cannot involve complete correspondence between p and any fact' (see above).

For Soames, then, there is no more to Frege's second argument than the statement of an undisputable, inconsequential assumption together with a brief justification ('since...'). As such it would neither be an argument against the correspondence theory nor contribute in any way to Soames's 'general argument' (see above), for as we shall see in section 10.2, Soames reconstructs the latter by means of a non-analytic definition of the form 'x is true:= x is T', so that it no longer includes any reference to correspondence at all. But this entirely misses Frege's original argument.

Soames's misunderstanding of Frege's second argument begins with the fact that he reads Frege as applying the notion of truth-as-correspondence to propositions – a misunderstanding that, once again,

is probably due to the influence of Tarski, who applied truth-as-correspondence to sentences (or seemed to do so). But Frege's claim that the
relata must be distinct clearly refers to the case of *ideas*:

> It would only be possible to compare an *idea* with a thing if the thing
> were an *idea* too. And then, if the first [*idea*] corresponds perfectly
> with the second [*idea*], they coincide. But this is not at all what peo-
> ple intend when they define truth as the correspondence of an *idea*
> with something real. For in this case it is essential precisely that the
> real thing be distinct from the *idea*. (v)

Thus, Frege's point is not, as Soames thinks, 'that the truth of [proposi-
tion] p cannot involve complete correspondence between p and any
fact.' Rather, Frege's claim that truth as correspondence must be imperfect (incomplete) –or, in other words, that the relata in a correspondence
relation must be distinct – pertains exclusively to I-truth. For scientific
truth, on the other hand, Frege claims just the opposite, namely *perfection* (completeness). Frege has first and foremost ideas (and pictures) in
mind when he says that

> it is essential precisely that the real thing be distinct from the idea.
> But then there can be no perfect correspondence, no perfect truth.
> So nothing at all would be true; for what is only half true is untrue.
> Truth does not admit of more or less. (v)

This passage makes it obvious that for Frege, the main point here is
that scientific truth is *perfect* (complete) truth not admitting of any
'more or less'. But this main thesis is nowhere reflected in Soames's
reconstruction.

Furthermore, we know that Frege has no problem with the identity
of (true) thoughts and facts. He explicitly states: 'What is a fact? A fact
is [!] a thought that is true' (Frege, 1918a, p. 74; see section 14.2). Thus,
there is no way that Frege would consider the following assumption as
undisputed with regard to thoughts (propositions), as Soames appears
to suggest: '[A] truth [p] is supposed to be distinct [!] from the fact it
corresponds to' (see above).

Like Soames, Künne refers to Frege's distinctness requirements only
affirmatively: '[Frege] correctly [!] assumed that it is essential for a correspondence theory that the 'correspondents' are distinct' (Künne, 2003,
p. 129). But Künne also does not comment on Frege's thesis that the
distinctness of the relata is incompatible with the perfection of truth!

In short: Soames and Künne miss the point of Frege's second argument because they overlook the crucial role of Frege's thesis that scientific truth must be perfect (complete).

5.4 Stuhlmann-Laeisz's critique

5.4.0 The only Frege scholar to thoroughly examine Frege's second argument is Stuhlmann-Laeisz (Stuhlmann-Laeisz, 1995, pp. 19–21). His understanding of Frege's second argument is similar to mine; yet unlike me he regards the argument as invalid. In what follows, I shall quote his critique of the argument and show that his critique is not sound. According to Stuhlmann-Laeisz Frege overlooks a certain logical distinction exemplified by the sentences 'Socrates is perfectly wise' and 'Socrates is perfect and wise'.

In his reconstruction of Frege's argument Stuhlmann-Laeisz makes 'use of the correlation f between a potential truth bearer [X] and its correlatum [f(X)]' (ibid.). According to Frege the deduction of

(5) 'X is perfectly true iff X perfectly corresponds to f(X)'

from

(1) 'X is true iff X corresponds to f(X)' (ibid.)

is correct. Stuhlmann-Laeisz' critique runs as follows:

> [...] the deduction of line (5) from line (1) presents a problem. Frege tacitly presupposes in his [...] objection that perfection is the same in both truths and correspondences. We would think that if the classes under consideration coincide – as the correspondence theory holds – then the class of perfect truths should also coincide with the class of X, that perfectly correspond to their respective f(X). But this presupposes that the statement 'X is perfectly true' is equivalent to the statement 'X is true and X is perfect', and this presupposition is not true. For 'perfect' is an *attributive adjective* whose connotation depends on its association with a certain noun. [...] Thus a perfect restoration of a building is perfect in a very different sense from a perfect workman's performance. This is why the fact that the restoration of a building is a workmen's performance does not establish the conclusion that a perfect restoration is a perfect workman's performance. For the same reason Frege's presupposition in his [...] objection

fails, namely that if truth is correspondence then perfect truth is perfect correspondence. Hence the argumentative move from (1) to (5) in the reconstructed argument contains a logical error. [ibid.]

5.4.1 The crucial point in Stuhlmann-Laeisz's argument is the concept 'attributive adjective'. Let us look at the word 'good' as another example for an attributive adjective and assume that the following equivalence is factually true:

> x is a 12th-grade student at Thomas Jefferson High School in 2011 [in short: a student] iff x is a player of the Thomas Jefferson High School baseball team in 2011 [in short: a player].

The predicates 'x is a student' and 'x is a player' have different senses, hence they are not synonymous; yet they (happen to) determine the same set of people. Now if we add the attributive adjective 'good' to these *non-synonymous* predicates, the resulting expressions 'x is a good student' and 'x is a good player' will usually not just differ in sense, but also determine different sets of people. However, Frege's proof – even in Stuhlmann-Laeisz's reconstruction – supposes a different kind of equivalence: that expressed in a *definition*, that is, when both sides of the equivalence are asserted to be *synonymous*:

> (1) X is true iff X corresponds to f(X).

Frege writes of the definiendum in an analytic definition: 'We believe that we can give a logical analysis of its sense, obtaining a complex expression which in our opinion has *the same sense*' (my emphasis; see the longer passage quoted in section 8.2). If (1) is a correct definition then we have two synonymous predicates on both sides of the equivalence. Hence, if we add the term 'perfectly'[1] on both sides of the equivalence, we again obtain two synonymous predicates. In particular, both predicates will determine the same set of entities. Thus, if (1) is correct as a definition then the following sentence is true as well:

> (5) X is perfectly true iff X perfectly corresponds to f(X).

However, since (5) is not true – something that Stuhlmann-Laeisz concedes – (1) cannot be an adequate definition: the two predicates on the left and right of the equivalence are not synonymous. Since a definition (something asserting synonymy) is at stake, Frege is justified in making the tacit assumption – which, according to Stuhlmann-Laeisz (1995, p. 21), is false – that perfection is the same in both truths and correspondences.

To date only Stuhlmann-Laeisz has provided a thorough analysis of the second argument, but his critique rests on the same logical error of which he accuses Frege, and so boomerangs right back at him.

6
The Third Argument: Scientific Truth Is Independent

6.1 The text

(vi) Or does it? [= Or does truth admit of more or less?] Could we not maintain that there is truth when there is correspondence in a certain respect? But which respect? And what would we then have to do so as to decide whether something were true? We should have to inquire whether it were true that, for example, an idea and something real correspond in the specified respect. And with that we should be confronted again by a question of the same kind, and the game could start all over.

[(ββ) Final conclusion of the refutations:] So *this attempt* [extending through (iv) – (vi)] to explain truth as correspondence [of ideas or pictures with something in reality] breaks down.

None of the interpretations known to me treats this paragraph as an *independent* argument in its own right; instead, following Dummett's influential work, it is generally interpreted as a variant or special case of the fourth argument (see Chapters 8 and 10). In either case, scholars follow Dummett's analysis of the argument's overall *structure*, according to which the third objection is intimately connected to the fourth. Geach and Stoothoff have considerably supported this reading by adding a dash after the second argument in their translation and by translating Frege's question 'Oder doch?' ['Or does it?'] only by 'but' (see appendix, section A.1).

On my interpretation, the common ground between the third and fourth arguments is limited to two points. First, both arguments share the presupposition that S-truth is, in a certain sense, omnipresent (or redundant) in propositional questions; moreover, Frege uses the stylistic tool of the propositional question only in both of these objections but not in the first two objections. Second, in contrast to the first two objections, both arguments criticize not the definiens of the definitions of truth discussed but rather their *applicability* in particular cases. Despite these similarities, however, the third argument is very different from the fourth. For, while the fourth argument is a circle objection (and is valid under Frege's premises), the third argument is, in my view, neither a circle objection nor a 'related' kind of objection such as a regress argument (see Stuhlmann-Laeisz, 1995, p. 66).

In fact, the third argument is much more closely connected to the first and second arguments than it is to the fourth. By means of the phrase 'in a certain respect' Frege adds the third component to his comprehensive definition of I-truth, whose three components are the topics of his first three arguments:

(D_{1-3}) I-truth is

(1) correspondence

(2) between an idea (a picture) and an object distinct from that idea (picture)

(3) in a certain respect H*.

Frege clearly separates the first three arguments from the fourth: *this* attempt (cf. $(\beta\beta)$) to define truth concerns I-truth as discussed in the first three arguments, while *any other* attempt concerns S-truth as discussed in the fourth argument (see sections 3.2 and 7.6). And while the first attempt fails for three content-related reasons (absoluteness, perfection, and independence), the second attempt fails due to a formal objection, namely, the circle objection.

I shall offer a comprehensive justification of my new interpretation of the third argument. In the present chapter I shall provide a close reading of the actual text of the third argument in 'Der Gedanke'. My justification also contains the following points, which will be discussed immediately following this chapter: Chapter 7 offers a detailed comparison of the passage in 'Der Gedanke' with a parallel textual passage in Frege's posthumously published fragment 'Logik' that also deals with the third and fourth objections. In this very similarly worded passage, Frege *explicitly* says that he sees the third argument as a special case of the fourth argument; this, of course, raises the question: why does he assess the connection between the two arguments differently in 'Der Gedanke'? Chapter 8 offers a novel reading of the fourth argument that is consistent with my interpetation of the third. Chapter 9 includes a clarification of Frege's presupposition that truth is, in some sense, omnipresent in propositional questions. Finally, in Chapter 10 I develop a detailed critique of Dummett's reading of the third and fourth argument as a regress argument.

6.2   The connection between the second and third arguments

Frege writes: 'Or does it? [= does truth admit of more or less?]. Could we not maintain that there is truth when there is correspondence in a certain respect?' (vi). If scientific truth is perfect then, if it is understood in terms of a correspondence, it must consist in perfect correspondence or correspondence in *every* respect. If, on the other hand, truth is imperfect and admits a 'more or less' (v), then this implies that one picture may be truer than another in the sense that the former corresponds to the depicted object in *more* respects than the latter. Thus, in this sense a picture of Cologne Cathedral that accurately depicts the Cathedral with regard to both shape and color may be truer than one that is accurate in only one of these respects.

The basis of this concept of truth as admitting a 'more or less' with regard to correspondence is the notion that in order to be true, a picture has to correspond to a depicted object *in a certain respect H**. It is this basic concept of truth as correspondence in a certain respect H* that becomes Frege's target in his third objection. And, as I shall argue now, by ruling out this concept of truth Frege at the same time succeeds in disposing of any concept of truth according to which truth admits of degrees – he disposes, that is, of the concept of a gradable or scalable form of truth altogether.

Suppose a picture is true to degree 1 if it corresponds to the depicted object in *one* respect, and that it is true to degree 2 if there are *two* such respects, ... , and it is true to degree n if there are n such respects. In this case we can say that truth of degree 5 constitutes a higher degree of truth than truth of degree 3. Now if we can show for each respect H, as Frege does in his third argument, that correspondence with respect to H has nothing to do with scientific truth or, in other words, that it would be nothing but truth of degree 0, we will have shown that correspondence in n respects does not constitute scientific truth. At most we could then say that it constitutes a scientific truth of degree 0, for $n \bullet 0 = 0$, but truth of degree 0 is still no truth at all.

I shall now present my reading of the third argument. In doing so, I start out with a reproduction of the actual text, numbering each relevant part for easier reference in the context of the following discussion.

6.3   The dependency of I-truth

6.3.0   Frege argues:

> (G1) Or does it? [= does truth 'admit of more or less' (v), after all?]
>
> (G2) Could we not maintain that there is truth when there is correspondence [of an idea or picture with something real] in a certain respect?
>
> (G3) But which respect?

Now, whatever respect we may choose in response to this question, there will be the follow-up ('then') question:[1]

> (G4a) And what would we then [= after stipulating the respect] have to do so as to decide
>
> (G4b) whether something were true?

Once the respect has been stipulated ('then'), that is, upon completion of the *definition* of I-truth, there is a yet further question (G4) about the *applicability* of that definition. I shall argue for this thesis in more detail in Chapter 9. At this point I restrict myself to presenting only two quotes in support of it. First, the parallel passage to (G4) in 'Logik' reads: '[...] in order to *apply* this [definition] we should have to decide in a particular case' (my emphasis, see section 7.1), and second, the corresponding passage in the fourth argument in 'Der Gedanke' reads: '[...] in *application* to any particular case it would always depend on [...]' ((vii), my emphasis).

The question about the *applicability* of a definition is important because even if a definition of I-truth can be established whose *definiens* holds up against all critique, we would still have to settle whether that definition of I-truth can be applied without problems in any particular case. In his third and fourth arguments Frege criticizes not the definiens but the applicability of the respective definitions, while his critique in the first and second arguments focuses on the definiens.

What, then, would we have to do in order to decide whether something is I-true in the sense of this definition? To illustrate how Frege answers this question, I shall first outline the standard way the question is addressed by proponents of a correspondence theory and then, in a second step, compare this standard way with Frege's way of addressing the question.

On the standard procedure, we would first replace the definiendum by the definiens. If we do this, we obtain the following sentence (which is included in, but does not exhaust, Frege's complete answer):

(G5*) We should have to inquire whether an idea and something real correspond in the specified respect.[2]

The subsequent standard steps in applying the definition would look like this: we should find the thing to which our respective idea refers and then conduct a comparison between the two relata. Depending on the results of this comparison, the idea would turn out to be either true or false, and with that our application of the definition to a particular case would be successfully completed without any further problems. This would be the standard procedure.

6.3.1   According to Frege, however, there is a problem in applying the definition of I-truth. His answer to question (G4) differs from answer (G5*) in that here he suddenly brings up the concept of S-truth:

(G5a) We should have to inquire

(G5b) whether it were [S-]true that, for example, an idea and something real correspond in the specified respect.

The decisive point in Frege's third argument is his question (G5b). Its first part, 'whether it were true', refers to the *thought* expressed by the second part of (G5b): 'that, for example, an idea and something real correspond in the specified respect.' This is why the term 'true' in the first part of (G5b) must refer to *S-truth*, while the second part of (G5b) contains the definiens of *I-truth* 'an idea and something real correspond in the specified respect'. Thus, question (G5b) employs altogether *two* concepts of truth; this can be illustrated more clearly if we look at it in the following manner:

(G5b) whether it were true [S-truth]
that, for example, an idea and something real correspond in the specified respect. [I-truth]

The appearance of S-truth in (G5b) comes as a surprise for the reader because in (G4b) Frege speaks exclusively of *I-truth*, as he does elsewhere in the first three arguments. Why is it that Frege all of a sudden decides to bring up S-truth in this context? This question will be eventually answered in section 6.4.

6.3.2   Frege comments as follows on the situation reached in (G5b):

(G6a) And with that [with (G5b)], we should be confronted again by a question of the same kind [...]

(G5b) is a 'question of the same kind' as (G4b). For in (G4b) we are confronted with the question of whether something, namely an *idea* or a picture (see section 2.6), is I-true in accordance with the third definition, and in (G5b) we are confronted with the question of whether a *thought* (or sentence) is S-true. Thus, (G4b) and (G5b) are questions of the same kind insofar as they both are questions about the truth of something.

We are first confronted with the question of whether something is *I-true*. In attempting to answer this question we end up being confronted with a question of the same kind, namely the question (G5b) of whether it is *S-true* that.... Thus, we can answer question (G4b) about the I-truth of an idea only if we can answer question (G5b) about the S-truth of a thought. Our attempt to define I-truth and to apply the concept in accordance with that definition cannot be completed independently of applying the competing concept of S-truth, to which we are referred via (G5b). In attempting to apply I-truth we notice that any such application is *dependent* on an application of the concept of S-truth.[3]

6.3.3   The second part of (G6) goes:

(G6b) [...] and the game could start all over.

Now, what game is it that we have played so far in the context of the third argument? In (G2)/(G3) we defined I-truth, and in (G4)/(G5) we looked into how to apply the concept thus defined. It is this game that could start all over if we now wish to answer the question raised in (G5b); only this time it would center around S-truth instead of I-truth: We should have to first define S-truth and then look into applying the concept thus defined.

It is important to understand that Frege does not speak of a *continuance* of the game, let alone of an *endless* continuance (a regress), but rather of a *restart*. To be sure, from the point of view of (G4) the restart of the game in terms of S-truth at the same time constitutes a continuance of the original game involving I-truth, since application of S-truth is a necessary component of applying I-truth. However, if we focus just on question (G5b) (see (G6a)) and wish to apply the concept of S-truth then we do not need I-truth to do so. The fact that in applying the definition of I-truth we replace the definiendum in (G4b) by the definiens in (G5b) does not contribute to the application of S-truth: for whatever content the second part of (G5b) has, the question of how to apply S-truth would still be the same.

Though according to (G6b) that game could start all over, Frege refrains from playing it again but breaks it off at this point. Obviously, he thinks that what he has shown up to this point suffices to establish that any attempt at applying the definition of I-truth must fail. This is at least what he says immediately following (G6).

6.3.4 Frege writes:

> (G7) So this attempt [extending through (iv) – (vi)] to explain truth as correspondence [of ideas or pictures with something in reality] breaks down.

Now, just why does the application of the definition of I-truth break down in the third argument? Let us recall why the definitions of I-truth in the first two arguments are rejected. The fact that I-truth is relational and imperfect does not disqualify I-truth as such but only as a candidate for scientific truth, since according to Frege the latter is absolute and perfect. A similar situation emerges in the third argument. Here, the very question

> (G5b) whether it were true [S-truth]
>
> that, for example, an idea and something real correspond in the specified respect. [I-truth]

shows that I-truth depends on S-truth in the process of applying the definition of I-truth. But here again, this dependence alone does not disqualify I-truth per se. For why should not one concept of truth depend on another one? But can, by contrast, truth in the *proper* sense of the word, namely 'that sort of truth which it is the aim of science to discern' (Frege, 1918a, p. 59), depend on any other concept of truth? The answer is: no. The dependence relation can work only one way, namely between an improper, secondary notion of truth, and the proper or primary notion of truth. Thus, it is precisely the dependence of I-truth on another, more fundamental concept of truth that disqualifies I-truth as a candidate for scientific truth (see Chapter 11).

6.3.5   This interpretation will strike anyone who has hitherto subscribed to Dummett's reading of Frege as outlandish. At the very least, the following questions will need to be addressed:

(1) What justifies Frege's insertion of the phrase 'whether it were true that' in (G5b)?
(2) What reasons speak against Dummett's interpretation according to which (G4b) and (G5b) are the beginning of an infinite regress?
(3) Is the strict separation of I-truth and S-truth in my presentation of the third argument justified?

I shall briefly address these questions, together with some other related issues, in the following sections, referring the reader on occasion to the chapters in which they are discussed in more detail. Concerning the 'road to the third argument', see also section 13.3.

6.3.6   Before I address the questions mentioned above, I should like to briefly present my view of the relation between definition, application, and dependency of concepts.

Application of a concept:[4] We apply a concept P by asking, about an object x from a suitable domain, whether x is P and providing an answer to this question (see (G4)). We apply concept P in accordance with a certain definition (= conceptual analysis) by using the concept in accordance with the sense specified in that definition. Thus, for example, we apply the concept of I-truth in accordance with Frege's third definition by asking whether idea I_0 of Cologne Cathedral is I-true in the sense of corresponding to that cathedral in a specified respect H*.

Application of a definition: Given the definition 'P:= A and B', the question of whether x is P and the question of whether x is A and B are identical (see section 6.4), since both questions contain the same concept P (= A and B). Both *questions* are identical, because both interrogative *sentences* are synonymous. Thus, whether we formulate our question as 'Is x P?', using the definiendum, or as 'Is x A and B?', using the definiens, does not matter in order to apply the *concept P/A and B*. In the second case, however, I speak of applying the *definition*, since in the interrogative sentence 'Is x A and B?' the definiendum 'P' has been replaced by the definiens.

Dependency of a concept: An application of the concept *P* depends on an application of a concept *Q* that is distinct from *P* if and only if the question of whether x is P can be answered only by answering (either previously or simultaneously) the question whether something (!) is Q. Thus, given our definition 'P:= A and B', an application of *P* depends on an application of *A* because the question of whether x is P can be answered only if we are able to answer the question whether x is A as well. The dependency of an application of I-truth from an application of S-truth is revealed in the fact that the question of whether, for example, our idea of Cologne Cathedral is I-true (that is, corresponds to that cathedral in a specified respect H*) can be answered only if we are able to answer the question of whether the thought that our idea of Cologne Cathedral is I-true (that is, ...) is S-true.

Dependency and definability: Most concepts are definable and are applied (if used correctly) in accordance with their definitions. But even if *Q* is an indefinable yet applicable concept the concept *P* may be dependent in its application in the sense described above on *Q*. As will be discussed in connection with Frege's fourth argument, Frege considers the concept of S-truth as indefinable. This does not diminish the dependency of I-truth on S-truth in applying the respective concepts.

6.4 The omnipresence of truth and Dummett's alleged regress

6.4.0 What justifies Frege's transition from

(G4b) whether something were true

to

(G5b) whether it were *[S-]true* that, for example, an idea and something real correspond in the specified respect?

To explain this transition in more detail, I shall begin by first splitting it up into two smaller steps. The transition from the question

(G4b) whether something were [I-]true?

to the question

> (G5b*) whether, for example, an idea and something real correspond in the specified respect?

is easy to follow under the condition that the concept of truth employed in (G4b) is that of I-truth. For in this transition we have merely replaced the definiendum of I-truth

> something [namely a picture or idea] *is true* (see (G4b))

by the definiens of Frege's third definition:

> an idea and something real correspond in the specified respect (see (G5b)).

Since definiendum and definiens are synonymous in an accurate definition, the interrogative sentences (G4b) and (G5b*) are synonymous as well. Thus, the transition from (G4b) to (G5b*) is unproblematic.

6.4.1   But Frege does not merely move on from (G4b) to (G5b*) but instead to the extended interrogative sentence

> (G5b) *whether it were true that*, for example, an idea and something real correspond in the specified respect.

Now the question

> (G5b) whether it were true that [...]

is an indirect question about the truth of a *sentence/thought*: the phrase 'whether it were true that...' requires the insertion of a sentence in place of the dots in order to become a complete indirect interrogative sentence. Thus, it seems obvious that the concept of truth employed in 'whether it were true that...' is that of S-truth. But how can Frege here all of a sudden introduce the concept of S-truth, given that what is at stake is exclusively the applicability of I-truth in accordance with the third definition?

According to Frege, S-truth is present in every assertoric sentence and every propositional question; it is in this sense *omnipresent* (Künne,

2003, pp. 35, 50ff). This presupposition of omnipresence can be briefly formulated as follows:

(F) The sentences 'p' and 'It is true that p' are synonymous.

(F*) The propositional questions 'p?' and 'Is it true that p?' are synonymous.

In deviation from standard linguistic terminology, I use the term 'synonymous' in the context of my Frege interpretation as applying whenever two sentences have the same *Fregean* sense, that is, whenever they express the *same thought*. This use of 'synonymous' does not consider the 'illumination' of a thought, which according to Frege can also be part of a sentence content but does not have any relevant role in his argumentation (see Frege, 1918a, p. 63).

According to (F*) the two questions (a) 'p?' and (b) 'Is it true that p?' are synonymous. This synonymy can be read in two directions. Moving from (a) to (b) we'll see that truth is *omnipresent*. Moving from (b) to (a) we'll see that truth is *redundant*. In Frege's arguments that are discussed here the point is always the step from (a) to (b), which is why I address omnipresence here instead of redundancy, as it is standardly done.

6.4.2   The first evidence that Frege claims (F) and (F*) can be found in the following passage of 'Der Gedanke': 'that the sentence "I smell the scent of violets" has just the same content as the sentence "it is true that I smell the scent of violets"' (Frege, 1918a, p. 61).

(F*) is likewise applicable to indirect propositional questions; thus, 'whether p' is synonymous to 'whether it is true that p'. This is why Frege's insertion of 'whether it were [S-]true that' in (G5b) is nothing but a reformulation of an interrogative sentence in terms of another, synonymous interrogative sentence and as such is justified under the presupposition of (F*). In the following I shall *make use* of (F*) in my reconstruction of Frege's third and fourth objections. I shall later *discuss* the omnipresence of truth in more detail in Chapter 9.

6.4.3   More evidence for the thesis that Frege considered (G4b) and (G5b) as synonymous comes from a parallel passage in 'Logik'. Here we read:

(L4a) [...] in order to apply this definition we should have to decide in a particular case

(L4b) whether an idea did correspond to reality,

(L5a) in other words:

(L5b) whether it is true that the idea corresponds to reality.[5]

Here, (L4b) and (L5b) express 'in other words' the same question; that is, the insertion of 'it is true that' does not change the (Fregean) sense of the interrogative sentence. Frege presupposes (F*) both in 'Logik' and in 'Der Gedanke'; his position on this matter does not change between the two writings.

6.4.4   Thus, Frege's grand transition from (G4b) to (G5b) can be divided into two smaller steps, each of which consists in a reformulation of an interrogative sentence in terms of another, synonymous interrogative sentence. In this way we get from the interrogative sentence

(G4b) whether something were [I-]true,

to

(G5b*) whether, for example, an idea and something real correspond in the specified respect,

by replacing the definiendum 'something is [I-]true' with the definiens of the third definition. This is the first reformulation in terms of a synonymous interrogative sentence. If we now apply (F*) to (G5b*) then we obtain, by way of the second reformulation in terms of synonyms, the interrogative sentence

(G5b) whether it were [S-]true that, for example, an idea and something real correspond in the specified respect.

According to Frege, we get from (G4b) to (G5b) by means of two transitional steps involving the replacement of synonyms, both of which are justified per se. This is why (G4b) and (G5b), which are *distinct* sentences with regard to wording, are synonymous, i.e. they are identical with regard to their content and therefore express *one and the same question*. Thus, we could not even ask question (G4b) about the truth of an idea without at the same time asking the corresponding question (G5b) about the truth of a sentence: by asking question (G4b) we simultaneously raise question (G5b) because both *questions* are identical even though the (synonymous) *interrogative sentences* are distinct.

I shall address the matter of omnipresence in more detail in Chapter 9.

6.4.5   Now what does speak against Dummett's interpretation, according to which Frege's third argument is a regress argument? According to Dummett, questions (G4b) and (G5b) constitute the beginning of an infinite regress that could be continued roughly as follows:

> whether it were true that it were true that, for example, an idea and something real correspond in the specified respect?

As evidence Dummett probably would cite the following sentence:

> (G6) And with that we should be confronted again by a question of the same kind, and the game could start all over.

(G4b) and (G5b) are the 'question[s] of the same kind' and are supposed to lead to an infinite regress. But a regress can arise only with numerically *distinct questions* of the same kind. (G4b) and (G5b), by contrast, are distinct but synonymous interrogative sentences and thus express the exact same question. But with *one and the same question* we cannot construct a regress, which is why (G4b) and (G5b) cannot possibly constitute the beginning of such regress. Furthermore, a regress is an endless game that starts *once* and thereafter continues ad infinitum; it is not a game that starts 'again', that is, a second time.

Künne actually notices as early as in 1985 that the two interrogative sentences (G4b) and (G5b) express the same question and that (F*) is not suited to generate an infinite regress precisely because it can serve to generate only synonymous sentences (see Künne, 1985, pp. 137–8). However, he uses this insight to *criticize* Frege based on his assumption (borrowed from Dummett) that Frege intended to construct an infinite regress. Yet, I would say that given the obvious incompatibility of a regress with (F*) and the assumption that Frege has a clear understanding of the concept of regress as well as of (F*), there is only *one* conclusion to be drawn with regard to the correct reading of Frege's argument: namely, that this argument is *not* intended as a regress argument.

I shall discuss Dummett's interpretation in more detail in Chapter 10.

6.5   The distinction between I-truth and S-truth

6.5.0   Now, is the strict distinction and comparison between I-truth and S-truth in my presentation of the third argument justified? Dummett and his followers do not make this distinction. Stuhlmann-Laeisz, for example, paraphrases Frege's sentences (G4)/(G5) as follows:

> To decide by means of such a definition whether a *sentence* is true we should have to determine whether it were true that that *sentence* corresponds to reality. And with this we are once again confronted with the question whether a *sentence* is true. (Stuhlmann-Laeisz, 1995, p. 25; my emphasis)

According to this reading, both (G4b) and (G5b) deal exclusively with the truth of *sentences*; no attention is paid to the difference between I-truth and S-truth. However, under appropriate scrutiny of the context, it hardly seems possible to ignore Frege's distinction of I-truth versus S-truth in his argumentation in the third and fourth paragraphs (see sections 2.5 and 2.6).

6.5.1   First, the main thesis of Frege's entire essay 'Der Gedanke' clearly shows where Frege locates his opponent: 'A thought belongs neither to my inner world as an idea, nor yet to the external world, the world of things perceptible by the senses' (Frege, 1918a, p. 75). Frege's opponent is psychologism, and he explicitly attacks the thesis that ideas are primary truth bearers. The proper truth bearers, namely thoughts, are not ideas. And this is why, as we may conclude, scientific truth cannot consist in I-truth but only in S-truth (under the assumption that Frege discusses only these two types of truth).

6.5.2   Second, Frege clearly distinguishes between I-truth and S-truth at the beginning of the third paragraph and then reduces I-truth to S-truth in the fourth paragraph:

> When we predicate truth of a picture we do not really mean to predicate a property which would belong to this picture altogether independently of other things. Rather, we always have in mind some totally different object and we want to say that that picture corresponds in some way to this object. 'My idea corresponds to Cologne Cathedral' is a sentence, and now it is a matter of the truth of this sentence. So what is improperly called the truth of pictures and ideas is reduced to the truth of sentences. (viii)

If Frege *explicitly* reduces the one concept of truth to the other, he must presuppose that they are different concepts. The relation between I-truth and S-truth is briefly indicated by Frege in the sentence:

> 'My idea corresponds to Cologne Cathedral [= this idea is I-true]' is a sentence, and now it is a matter of the [S-]truth of this sentence. (viii)

Precisely the same kind of relation is involved in question (G5b):

> (G5b) whether it were [S-]true that, for example, an idea and something real correspond in the specified respect [= the idea is I-true].

According to both quoted passages, I-truth consists in correspondence and holds only if the sentence in which that correspondence between an idea and the corresponding object in reality is expressed is S-true. Thus, we determine whether the idea is I-true by examining the S-truth of the corresponding sentence. And it is precisely in this sense that the application of I-truth depends on an application of S-truth. The *dependency* of I-truth on S-truth exposed in (G5b) precisely corresponds to the *reduction* of I-truth to S-truth that takes place in the fourth paragraph.

6.5.3   Third, a comparison with the parallel passage in 'Logik' also supports the strict distinction between I-truth and S-truth in (G5b).

Frege takes many of his arguments in 'Der Gedanke', including the third and fourth arguments, from his piece from twenty years earlier, 'Logik'. He does, however, considerably alter these two arguments in the later essay. And in my view, these changes can be properly understood only under the assumption that Frege comes to distinguish more strictly between I-truth and S-truth in 'Der Gedanke' than he had done in the parallel passage in 'Logik'.

I shall provide a more detailed comparison of the two texts in Chapter 7.

6.6   The independence of scientific truth

The crucial point of the third argument is that the dependency of I-truth on S-truth with regard to their application is revealed in

> (G5b) whether it were true that, for example, an idea and something real correspond in the specified respect.

This dependency does not disqualify the concept of I-truth in itself, but only the claim that I-truth is to be regarded as the original, scientific kind of truth. As the fourth paragraph clearly states, I-truth is nothing but a secondary, 'improper' kind of truth that can be 'reduced to the truth of sentences'. In the fourth paragraph Frege presents the sentence

(a) My idea corresponds to Cologne Cathedral

as a *meaningful* sentence and has no problems with it as such. Objectively speaking, there is no reason not to call the idea I-true if we do so *because* (a) is a true sentence, as long as we do not mistake I-truth for proper truth. However, such use of language will inherently always bear the risk of possible mix-up and misuse.

In short: if we apply I-truth in accordance with the third definition then (G5b) reveals its *dependency* on S-truth. Since, however, scientific truth is independent, it cannot be the same as I-truth. This reconstruction of the third argument meshes well with the first two objections: a scientific truth that is *absolute* and *perfect* cannot depend in its application on another kind of truth but is *independent* as well.

6.7   The game and its potential restart

(G6b) [...] and the game could start all over.

6.7.0   Frege does not elaborate on the game that he says could start all over. Hence, any clarification of that game's nature inevitably goes beyond the actual text. What I said concerning that game in section 6.3 above refers only to the context of the third argument. If we broaden our perspective to include the entire context of the third and fourth paragraphs, we may come to the following considerations.

Up until this point the game played in the third paragraph consists in answering the following questions:

(I) (a) Are *ideas* the truth bearers?

(b) How can *I-truth* be defined and

(c) how can the respective definition be applied?

(d) Is I-truth scientific truth?

Questions (b) and (c) can be answered in the affirmative, since definitions (D$_1$) – (D$_3$) together constitute a positive answer to (b), and its application (c) is possible in dependence on S-truth. Question (d), however, is answered in the negative because scientific truth is absolute, perfect and independent while I-truth is relative, imperfect and dependent according to (b) and (c). Thus, the answer to (d) follows from the answers given to (b) and (c). And from (d) follows the answer to (a): since the truth of ideas does not qualify as scientific truth, ideas cannot be the bearers of scientific truth. With this, the attempt to determine ideas as bearers of scientific truth, and I-truth itself as scientific truth, has failed.

And now the game of identifying the nature of scientific truth and its bearers could start 'all over', that is, from the beginning. And in analogy to the first run, this restart of the game would concern not I-truth and ideas but rather S-truth and sentences/thoughts as primary bearers of truth:

(S) (a) Are *sentences/thoughts* the truth bearers?

(b) How can *S-truth* be defined and

(c) how can the respective definition be applied?

(d) Is S-truth scientific truth?

Indeed, it is conceivable that the game *could* start all over at this point with the question of how S-truth could be appropriately defined, and so on. Yet Frege refrains from restarting the game since, after all, the starting point in this run is different from the one he faced in the case of I-truth. For in the case of S-truth, questions (a) and (d) can be answered *indirectly* already at this point even though Frege has not yet given any definition of S-truth: since only I-truth and S-truth are offered as candidates for scientific truth and I-truth has been disqualified, S-truth is the only remaining candidate for scientific truth, and sentences/thoughts must be the primary truth bearers. Thus, in order to give an affirmative answer to (a) and (d) we do not need to have an affirmative answer to (b) and (c) in this case. And there is no affirmative answer to (b) and (c), anyway; instead, questions (b) and (c) are subsequently answered in the negative via Frege's later circle objection: any application of an analytic definition of S-truth fails due to a circle.

6.7.1 Frege's game in the third paragraph up to this point serves to answer *two* closely related questions: on one hand, the question about the nature of truth *bearers* and, on the other hand, the right (analytic) *definition* of truth. The result of the first three arguments concerning I-truth can be summarized as follows: the *truth of ideas* is not scientific

(= proper) truth. This result can be collectively applied to both questions and is supplemented, in other locations in the text, by the corresponding findings concerning S-truth.

First, ideas cannot be the proper *bearers* of truth because the truth of ideas is not truth in the proper sense of the word. Hence, only the sentence or thought remains as potential bearer of scientific truth. The topic of truth bearers is taken up again in the fourth paragraph. There I-truth is reduced first to the S-truth of sentences and eventually to the S-truth of thoughts; with that, S-truth is shown to be truth in the proper sense of the word and the thought to be the proper bearer of truth.

Second, scientific truth cannot be *defined* as I-truth. The topic of truth's definability is continued in the fourth argument (contained in the third paragraph). There, it is shown that scientific truth cannot be defined as S-truth, either, because every (analytic) definition of S-truth fails due to a circle in its application. Thus, it follows from the four arguments that scientific truth is in principle indefinable (under the condition that only I-truth and S-truth are considered to be possible candidates).

The fourth argument contributes only to the question of the definability of S-truth but not to that of the proper (primary) bearer of truth. Conversely, the fourth paragraph does not contribute to the question of definability but only to the question of proper truth bearers. Thus, with regard to S-truth the two questions are answered in part separately, while with regard to I-truth their answers are closely connected in the first three arguments.

6.8   The connections between the first three arguments

6.8.0   Frege's three objections against the three offered definitions of I-truth (truth of ideas/pictures) are each based on one of three essential characteristics of scientific truth, namely, *absoluteness*, *perfection*, and *independence*. Though these characteristics do not suffice to define scientific truth, they do suffice to rule out I-truth as scientific truth. Since there is no third candidate, it follows that S-truth remains as the only adequate candidate for scientific truth. I shall now illustrate the interconnections between the three arguments in three different ways.

6.8.1   Comparing the three *objections* in these arguments, we notice that Frege seemingly makes increasing concessions to I-truth as correspondence.

According to the first argument, scientific truth is an *absolute* and therefore, in principle, not a *relative* (or relational) truth. Since I-truth in terms of correspondence is relative truth, it is ruled out as scientific truth.

Yet for those who assume that scientific truth is in fact correspondence, the second argument shows that, in that case, it would require perfect correspondence in order to be *perfect* truth, which would involve a correspondence in every respect and hence numerical identity of the relata. For I-truth, by contrast, it is essential that the idea is not identical to what it depicts, and thus in I-truth the correspondence must *not* hold in every respect. It follows that I-truth is not perfect truth and is therefore unsuitable as a candidate for scientific truth.

Again if we assume that scientific truth may even be imperfect (admitting of more or less), then it may consist in correspondence *in a certain respect*. But as the third argument subsequently shows, since I-truth depends on S-truth while scientific truth is independent, I-truth is once more ruled out as a possible candidate for scientific truth. What is peculiar about the third argument is that it doesn't just rule out I-truth as scientific truth, but that it does so by means of the concept of S-truth, while S-truth is not even mentioned in the first and second arguments.

6.8.2   If we compare the three *definitions* (D_1), (D_2), and (D_3) of I-truth we will notice that they are connected to one another just like a series of consecutive specifications of that concept.

(D_1) Idea x is true:= Idea x *corresponds to* the depicted object.

(D_2) Idea x is true:= Idea x corresponds to the depicted object, which is *distinct from* x.

(D_3) Idea x is true:= Idea x corresponds to the depicted object *in a certain respect H**.

We can combine the three definitions (D_1), (D_2), and (D_3) into *one* comprehensive definition of I-truth:

(D_{1-3}) I-truth is

(1) Correspondence

(2) between an idea and an object distinct from that idea

(3) in a certain respect H*.

We can regard this definition as a plausible, adequate, and perhaps even *the* 'natural' definition of truth as correspondence for ideas and pictures. Frege's first three arguments serve to attack the *three* components of this *one* definition.

6.8.3   Corresponding to the comprehensive definition above, we can combine Frege's first three arguments into one comprehensive argument as follows:

(1) I-truth consists in a correspondence between an idea and a numerically distinct object in a certain respect H*.
(2) If I-truth consists in a correspondence then it is relative (relational).
(3) If I-truth consists in a correspondence between numerically distinct relata then it is imperfect.
(4) If I-truth consists in a correspondence between numerically distinct relata in a certain respect H* then it is dependent.
(5) But scientific truth is absolute, perfect, and independent.
(6) Therefore (for three substantial reasons), I-truth cannot be scientific truth.

6.8.4   With respect to (4) it can be easily seen that the dependency of I-truth is not a consequence specifically of the *certain respect H*￪* in which the correspondence is to hold. At first glance, this seems different with regard to (2) and (3): here, the correspondence in (2) by itself seems to imply the feature of relativity, and the difference between the relata in (3) implies imperfection.

But in fact, even in (2) and (3) the 'then' clauses are not *specific* consequences of the 'if' clauses. For each of the three 'then' sentences in (2), (3), and (4) would follow from *either* of the corresponding 'if' sentences, not just from the one immediately preceding them (cf. section 5.2 with respect to (2) and (3)). Thus, we could also say:

(2') If I-truth consists in correspondence then it is relational, imperfect, and dependent (you can easily show that all three consequences follow).

The same can be said about (3) and (4). Though each of the three definitions adds on to the previous one(s), the three arguments connected to the 'then' clauses in (2) – (4) can be used against all three definitions.

6.8.5 Frege himself emphasizes the connections between his three arguments by clearly indicating the beginning and the end of his refutation of the correspondence theory (of I-truth). At the beginning stands the assumption to be disproven indirectly:

[(β)] It might be supposed from this that truth consists in a correspondence of a picture to what it depicts. (iv)

And directly following the third argument Frege formulates the result of his argumentation thus:

[(ββ)] So this attempt to explain truth as correspondence breaks down. (vi)

This attempt of which Frege speaks here concerns, in a word, definition (D_{1-3}); it extends over all three components of this definition. In other words: this attempt covers *all* of the three supplemental definitions (D_1), (D_2), and (D_3) that Frege discusses in his three arguments.

7
Parallels in Frege's 'Logik'

7.1 Similarities between the parallel arguments

7.1.0   In 1897, a good twenty years prior to 'Der Gedanke', Frege writes a piece titled 'Logik', which will remain a fragment to be subsequently first published only in his *Posthumous Writings*. Early in the text we come across a passage that is *very* similar to the third and fourth arguments in 'Der Gedanke'. As before, my numbering of the arguments as the third and fourth, respectively, refers to the order in which the arguments are presented in 'Der Gedanke' even when I discuss the parallel passages in 'Logik', where there is no first or second argument prior to the 'third' one.

A little later in 'Logik' we encounter another parallel passage that is somewhat *less* similar: the so-called treadmill argument, which some contemporary scholars treat as a paraphrase of the third and others as a paraphrase of the fourth argument. In the present chapter I shall thoroughly compare the first parallel passage in 'Logik' with the corresponding passage in 'Der Gedanke'; I shall also discuss the treadmill argument in section 7.9.

The text of the first parallel passage in 'Logik' does not leave any doubt that Frege intends both the third and the fourth arguments to proceed by showing the existence of a circle. When comparing this passage to the corresponding passage in 'Der Gedanke', the most striking change is that the circle objection in the third argument in 'Logik':

> (L6) Thus we should have to presuppose the very thing that is defined

is replaced by the following sentence in 'Der Gedanke':

(G6) And with that we should be confronted again by a question of the same kind, and the game could start all over.

Since Dummett this sentence has generally been read as if Frege here indicated an infinite regress. But if this were correct, Frege's alleged transition from a circle objection to a regress objection would have to be justifiable based on the remaining textual changes – if the transition from the earlier to the later text can be rationally reconstructed at all. But, as I shall argue in the following, the changes that Frege in fact makes in this argumentation do not support the presence of a regress argument but, rather, my interpretation of the third objection.

7.1.1 At first sight, the similarities between the first parallel passage in 'Logik' and the third and fourth arguments in 'Der Gedanke' are so overwhelming that both texts are often considered as presentations of *one and the same* argument. Thus, Dirk Greimann writes,

> There are two formulations of this argument [i.e., the argument for 'Frege's thesis of the indefinability of truth'] in Frege's entire work. They are as follows:
>
> [Here Greimann quotes the parallel passages in 'Logik' and 'Der Gedanke', respectively.][1]
>
> In both cases, Frege's argumentation for his indefinability thesis comes in the form of a *reductio ad absurdum* – the counter-assumption is shown to result in a contradiction. However, Frege's formulations do not make it clear whether the contradiction is derived by way of a circularity claim ('we should be going around in a circle') or a regress claim ('the game could start all over'). (Greimann, 1994, p. 78)

The following claims in Greimann's exposition are important, especially with regard to the third argument in 'Der Gedanke'.

First, the arguments in the passage from 'Logik' (hereinafter L) and in its parallel passage from 'Der Gedanke' (hereinafter G) quoted further below are essentially identical, because they are but 'two formulations of this [that is, the same] argument'.

Second, the two corresponding passages, and especially the third argument in both passages, have the same argumentative goal, namely the 'indefinability thesis'.

Third, the formulation 'das Spiel könnte von neuem beginnen' ['the game could start all over'] by itself is sufficient evidence that Frege here makes a 'regress claim'.

Fourth, 'Frege's formulations do not make it clear' whether he intends to use his entire argumentation, which includes the third and the fourth arguments, to derive a circle (as indicated in 'we should be going around in a circle') or a regress (as indicated in 'the game could start all over').

I believe all four of these claims to be erroneous.

7.1.2   In the following, I shall thoroughly compare the two text passages which are shown in Table 7.1. For this purpose I shall align sentences (L1) – (L12) from the 'Logik' text with their corresponding sentences (G1) – (G12) from 'Der Gedanke'. (The numbers in parentheses are not part of the text quoted, but my additions.) The sentences are thus arranged in two columns, one for each text, in such a way that – to the extent that the original order of the sentences is maintained – corresponding sentences/phrases from the two texts are opposite one another in the same row. The most important differences are indicated by italics in each case. Whenever I shall subsequently refer to the entire texts consisting of (L1) – (L12) and (G1) – (G12), respectively, I shall use the abbreviations L and G.

Like Künne and Soames, Greimann quotes the third argument in 'Der Gedanke' from (G2) onwards; however, it actually starts one question earlier, namely with (G1). The 'not' in (G2) makes it clear that (G2) is intended as an answer to question (G1), which is one reason why we should regard (G1) as belonging to the third argument as well. Another reason is that the sentence immediately preceding (G1), 'Truth does not admit of more or less', is much better suited to concluding the second argument, in which the perfection of truth is asserted, than (G1) is. Finally, (G1) illustrates how closely connected the third argument is to the second one.

Given these considerations, if we quote the third argument only from (G2) onwards – as is common – then we make the mistake of taking it out of its connection with the first two arguments, with whom it is much closer connected than with the fourth one. In *Conceptions of Truth* (p. 129) Künne even quotes the third argument from (G4) onwards.[2]

7.1.3   When comparing the two texts, we at once notice very broad similarities: Both L and G are structured into two arguments, and the

Table 7.1 The third and fourth arguments in 'Logik' and 'Der Gedanke'

(L1) *Now it would be futile to employ a definition in order to make it clearer what is to be understood by 'true'.*	(G1) *Or does it? [=does truth admit of more or less?]*
(L2) If, for example, we wished to say: 'An idea is true if it corresponds to reality',	(G2) Could we not maintain that there is truth when there is correspondence *in a certain respect?*
(L3) nothing would have been achieved, since	(G3) *But which respect?*
(L4a) in order to apply this definition we should have to decide in a particular case	(G4a) And what would we then have to do so as to decide
(L4b) whether an idea did correspond to reality,	(G4b) whether something were true?
(L5a) or in other words:	(G5a) We should have to inquire
(L5b) whether it is true that the idea corresponds to reality.	(G5b) whether it were true that, for example, an idea and something real correspond in the specified respect.
(L6) *Thus we should have to presuppose the very thing that is defined.*	(G6a) *And with that we should be confronted again by a question of the same kind,*
	(G6b) *and the game could start all over.*
(L7) *The same would hold of any definition of the form:*	(G7) *So this attempt to explain truth as correspondence [as developed in the three definitions] breaks down.*
(L8) [See (L7).]	(G8) *But likewise, any other attempt to define truth also breaks down.*
(L9) 'A is true if and only if it has such-and-such properties or stands in such-and-such a relation to such-and-such a thing'.	(G9) For in a definition certain characteristics would have to be specified.
(L10a) In each case in hand it would always come back to the question	(G10a) And in application to any particular case it would always depend on
(L10b) whether it is true that *A* has such-and-such properties, or stands in such-and-such a relation to such-and-such a thing.	(G10b) whether it were true that the characteristics were present.
(L11) [see (L6) and (L7).]	(G11) *So we should be going round in a circle.*
(L12) Truth is obviously something so primitive and simple that it is not possible to reduce it to anything still simpler.	(G12) So it is likely that the content of the word 'true' is sui generis and indefinable.

(Frege, 1897, p. 139 f (128 f))

sentences crucial for the two arguments (L2) – (L5)/(L9) – (L10) and (G2) – (G5b)/(G9) – (G10) are by and large identical. Even the conclusions (L12) and (G12) are identical in their essential claim: that truth is indefinable (whether this claim be 'obvious', as asserted in L, or merely 'likely', as asserted in G).

Now in the light of such an overwhelming similarity between the two texts, can we not ignore the seemingly minor differences and simply follow Greimann in speaking of 'two formulations of this [that is, the same] argument'? In what follows I argue that the differences between the two parallel passages are greater than they appear at first glance.

7.2   The connection between the two arguments

7.2.0   The most important difference between the two texts consists in the way in which Frege describes the connection between the third and fourth arguments.

In L *both* definitions are subject to the *same* kind of failure in their applications, namely a circle. There are three reasons for this thesis.

First, according to Frege's *explicit claim* in (L6), the definition in the third argument fails due to a circle, and in (L7) he again *explicitly* asserts the same with respect to the definition given in the fourth argument:

(L6) Thus we should have to presuppose the very thing that is defined.

(L7) The same would hold of any definition of the form: ...

Second, the *transition* from the definition

(L2) [...] An idea is true if it corresponds to reality

to

(L5b) whether it is true that the idea corresponds to reality,

precisely mirrors the transition from the definition

(L9) *A* is true if and only if it has such-and-such properties or stands in such-and-such a relation to such-and-such a thing

to

(L10b) whether it is true that *A* has such-and-such properties, or stands in such-and-such a relation to such-and-such a thing.

Both cases exemplify a transition from a definition of the form

    x is true:= x is T

to a clause of the form

    whether it is true that x is T.

If this transition constitutes a circle in the third argument then it ought to do so in the fourth argument as well.

    Third, the definition

    (L2) An idea is true if it corresponds to reality

is a *special case* of the definition

    (L9) *A* is true if and only if it has such-and-such properties or stands in such-and-such a relation to such-and-such a thing,

for 'corresponding to reality' does look like a special case of 'standing in such-and-such a relation to such-and-such a thing'. It follows that this special case fails for the same reason as the general definition.

    In short: there are three good reasons why *both definitions* in L fail *in the same way* in their application, namely, due to a circle. And this is the connection between the two arguments in L, both according to Frege's explicit *statements* and to the way he *proceeds*.

7.2.1 The connection between the two arguments looks very different and rather obscure in G. To illustrate this, I shall here run through the three reasons in the order in which they are stated above and also discuss the two arguments in L in connection with the second and third reasons.

    First, Frege's explicit claims about the reasons why the definitions fail *differ*. While Frege maintains the circle objection for the fourth definition in G,

    (G11) So we should be going round in a circle,

he obviously refrains from doing so for the third argument, where his objection is different:

    (G6) And with that we should be confronted again by a question of the same kind, and the game could start all over.

According to Frege's explicit claims the circle objection here pertains only to applications of the fourth definition but not to those of the third one.

Second, in contrast to Frege's explicit claim, the transition from the third definition in G to sentence (G5b), which is *not* subject to a circle objection, is *exactly the same* as in those three other cases in which Frege does speak of a circle. Compare the following clauses:

(G5b) whether it is true that, for example, an idea and something real correspond in the specified respect.

(G10b) whether it were true that the characteristics were present.

(L5b) whether it is true that the idea corresponds to reality.

(L10b) whether it is true that *A* has such-and-such properties, or stands in such-and-such a relation to such-and-such a thing.

All four of these cases deal with an application of a definition of the form

x is true:= x is T,

which leads to a question of the apparently problematic form

whether it is true that x is T.

In all of these four cases we obtain the problematic question at stake by applying (F*) and adding 'whether it were/is true'. Since the basic structure is identical in all four cases, we should expect that the circle objection is raised in all of them, and not just in three of them. The similarity of the manner in which Frege *proceeds* seems to contradict what Frege explicitly *says* in (G6).

Third, we should look into whether the definition used in (G5b) is a special case of the one used in (G10b). If we interpret the expression 'characteristic' (a) *broadly* then (G5b) is a special case of (G10b), but if we interpret 'characteristic' (b) *narrowly* then it is not.[3] Option (a) would be inconsistent with a special status for (G5b), while option (b) would be consistent with it. Thus, in the conflict between the first two reasons, the third one remains neutral.

7.2.2   In L there are *three* good reasons to assume that a circle objection is made both in the third and the fourth arguments. In G, however, there remains only *one* reason in favor of the claim that the circle objection is present in both arguments: the manner in which Frege argumentatively *proceeds* in all four cases is so similar that if there is a circle in three of them there ought to be one in the fourth. This raises the question: what has changed in the third argument from L to G so that it is no longer a circle objection?

Up to this point, all Frege scholars have interpreted the sentence

(G6) And with that we should be confronted again with a question of the same kind, and the game could start all over

as pointing out an infinite regress as the continuation of questions (G4b) and (G5b). But whoever continues to claim that (G6) is a regress objection ought to explain why Frege, even though he proceeds in the same manner four times (see above), speaks of a circle only three times and means (without speaking of) a regress the fourth time. Based on the changes leading from (L2) – (L5b) to (G2) – (G5b), it would have to be explained just why it no longer is supposed to be a circle objection but a regress objection.

Now, there may be many reasons why Frege does not repeat the circle objection from the L-version in the G-version of the third argument. Some of these reasons may be consistent with the assumption that at the time of 'Der Gedanke' Frege still regards the circle objection as valid. Nonetheless, in what follows I assume that Frege abandons the circle objection from 'Logik' because for a certain reason he comes to regard that objection as *invalid*, and that this reason can be disclosed by analyzing the texts. This assumption proves to be very constructive in that it can explain many differences between L and G.

7.3 Differences between the definitions

7.3.0 At first sight, the definitions in the fourth argument of L and G, respectively, clearly differ from one another:

(L9) *A* is true if and only if it

[(a)] has such-and-such properties or

[(b)] stands in such-and-such a relation to such-and-such a thing.

(G9) For in a definition certain characteristics would have to be specified.

Both definitions are analytic definitions.[4] But to what extent do they coincide with regard to content? (L9) covers all analytic definitions of the following structure. The definiens contains

(a') at least two one-place predicates interconnected by 'and', or

(b') at least one two-place predicate with suitable singular terms.

Since all analytic definitions are of either the (a') or the (b') type structure, (L9) covers the structure of *all* possible analytic definitions of truth.

Now, what are those 'properties' mentioned in (L9a) or the 'one-place predicates' mentioned in (a')? There are two possible interpretations:

(i) The properties are absolute or relational properties; that is, the one-place predicates could also include relational predicates (cf., 'x is true:= ∃y (x corresponds to y)'. This is the *broad* interpretation.

(ii) The properties are absolute properties; that is, the one-place predicates do not include relational predicates. This is the *narrow* interpretation.

(i) is supported by the fact that *relational* properties such as being a wife are perfectly common for modern logicians and are therefore considered a type of property. But if (i) were the correct interpretation, (L9) would be redundant; the relational property of 'stand[ing] in such-and-such a relation to such-and-such a thing' (see (L9b)) would be implied by the expression 'has such-and-such properties' (see (L9a)).

Accordingly, (ii) is supported by the fact that under it (L9) is *not* redundant. It is conceivable that Frege here uses the term 'properties' not in the sense of modern logic but rather as synonymous with 'absolute properties'. When Frege says in 'Der Gedanke' that the word 'true' appears as an adjective (or 'property-word', as the literal translation of 'Eigenschaftswort' would be) as opposed to a relation-word, he means by this that truth is not a relational property (see sections 4.2 and 13.2). This terminology would be consistent with the assumption that Frege in (L9) defines truth by means of (a) absolute properties or (b) relational properties.

Whether (L9) is redundant or not, either way (L9) covers all possible definitions of truth.

In the third argument of L the truth bearer is an idea, while it is designated as 'A' and left unspecified in the fourth argument of L. I understand 'A' here to stand for *any* truth bearer, whether it be an idea, a picture, a sentence, or a thought.

Overall, we can say with regard to (L9) that it includes all possible definitions of truth both with regard to their truth bearers and to their formal structure.

7.3.1   (G9) is also susceptible to two interpretations, which are analogous to the two interpretations of (L9) just discussed:

(i) In the *broad* interpretation, the 'certain characteristics' in (G9) include both absolute and relative properties.

(ii) In the *narrow* interpretation, the aforementioned 'characteristics' include only absolute, that is, non-relational properties.

As in the case of (L9), (i) is supported by our contemporary understanding of properties. If we then also admitted all kinds of truth bearers, (G9) would constitute a definition that is as comprehensive as (L9). However, we would have to accept a consequence that eventually undermines (i): according to the reading suggested by (i), the definition in (G2)/(G5) is a special case of (G9); hence the circle objection against (G9) would also apply to the definition offered in the third argument. Since Frege, however, abandons the circle objection in the third argument, the third definition cannot be a special case of the fourth one.[5] Thus, (G9) cannot cover all possible definitions of truth, as (L9) does, but must be restricted to a narrower scope of application. (G9)'s scope could be narrowed compared to that of (L9) by restricting the former so as to not allow for ideas as truth bearers or by adopting the narrow interpretation of 'characteristics'.

Apart from the factors that count against (i), a positive reason in favor of (ii) is the way Frege distinguishes between (G9) and the immediately preceding attempt at a definition:

(G7) So this attempt to explain truth as correspondence [comprising definitions (D1) – (D3)] breaks down.

(G8) But, likewise, any other attempt to define truth also breaks down.

This attempt (see (G7)) consists in defining truth as a relational property, namely as a *correspondence* of pictures or ideas. Since Frege discusses only *two* ways to define scientific truth in the third and fourth paragraphs, namely as either S-truth or I-truth, *any other* attempt (see (G8)) presumably refers to defining scientific truth as S-truth.

Frege's first argument already establishes that *any* theory of relational truth is unsatisfactory simply because it would always be incompatible with the absoluteness of scientific truth. This is why only one possible type of truth definition remains to be examined after (G8): can the truth of sentences/thoughts be defined as absolute truth? If so, this would be possible only by means of absolute properties, that is, 'characteristics' as conceived in the narrow interpretation of (ii).

In any case, it seems obvious that the definition in (G9) cannot be read as broadly as the one offered in (L9). Whether the required restriction lies with the nature of the truth bearers or with the so-called 'characteristics', or with both of them, remains to be seen.

7.3.2   There are also differences between the two respective L and G versions of the definition offered in the third argument. According to (L2) and (G2)/(G5), respectively, we can summarize these versions as follows:

(L2*) An idea x is true:= x corresponds to reality.

(G2*) An idea x is true:= x corresponds to something real in a certain respect.

(G2*) is more precise than (L2*) in two respects:

(1) The yet unspecified correspondence relation 'x corresponds to y' becomes 'x corresponds to y *in a certain* respect'.

(2) The second relatum is no longer *reality* in general but *something real* (namely some particular real thing).

If a picture of Cologne Cathedral – which is Frege's example – is true, this is not because it corresponds to *all of* reality, but because it corresponds with this *particular* real object, Cologne Cathedral. Since the correspondence cannot hold in *every* respect (otherwise the picture and the cathedral would be numerically identical), it can only hold in a *certain* respect. Both changes are unproblematic improvements of the earlier definition and entirely in sync with standard correspondence theory.

Can these changes, however, lead Frege to abandon the circle objection in the third argument and to replace it with a regress objection? Or is there another reason why Frege abandons the circle objection in relation to the third definition?

7.3.3   If we compare the two definitions in the third and fourth argument in L, namely

> (L2) An idea is true if it corresponds to reality,

and

> (L9) *A* is true if and only if it has such-and-such properties or stands in such-and-such a relation to such-and-such a thing,

then we notice that (L2) is a special case of (L9) (see section 7.2). Frege accordingly connects the two arguments as follows:

> (L7) The same would hold of any definition of the form: ...

Immediately after this, Frege offers definition (L9), which is a generalization of (L2).

But in G the connection between the arguments is established in a different way. For here, Frege follows up his critique of 'this attempt' (in German: 'dieser Versuch') with a critique of *any other attempt* (see (G7)/(G8). Since Frege strictly distinguishes in the third and fourth paragraphs between S-truth and I-truth and reduces the latter to the former, it is obvious that I-truth in G is different from, and not a special case of, S-truth. The definition

> (D₃) Idea x is true:= x corresponds to the depicted object in a certain respect H*

is *not* a special case of the definition

> (D₄) *Sentence/Thought* p is true:= p is A and p is B.

Looked at from this angle, it is conceivable that the circle objection Frege uses in the L text to attack (L9) is also intended to bring down its *special case* (L2). Definition (D₃) and the (very *different*) attempt (D₄) in G, by contrast, are confronted by Frege with different objections.

7.3.4   Comparing the third definition in G

> (D₃) x is true:= x corresponds to the depicted object in a certain respect H*

with the fourth definition in L

> (L9) *A* is true if and only if it has such-and-such properties or stands in such-and-such a relation to such-and-such a thing,

we notice at once that (D$_3$) is a special case of (L9), as *A* here stands for any kind of truth bearer. Thus, if the circle objection works against (L9), it also works against (D$_3$). The reverse can be said as well: if the circle objection fails in application to (D$_3$) then it also fails with respect to (L9). In this case the changes that Frege introduces in getting from (L9) to (G9) would have to explain why he abandons the circle objection with regard to (D$_3$) – and thereby also with regard to (L9) – while upholding it for (G9).

7.4   Proof structure of the arguments

7.4.0   The *starting point* of both the third and fourth arguments in L and G, respectively, consists in a definition of truth, while the end point is in each case one of the following sentences:

> (L5b) whether it is true
>> that the idea corresponds to reality.
> (L10b) whether it is true
>> that *A* has such-and-such properties, or stands in such-and-such a relation to such-and-such a thing.
> (G5b) whether it were true
>> that, for example, an idea and something real correspond in the specified respect.
> (G10b) whether it were true
>> that the characteristics were present.

In each of these sentences it is obvious for Frege that the application of the respective definition has failed: in three of the cases, the failure is traced to a circle, but in the fourth case, apparently, to something else. But how does Frege arrive at these sentences?

7.4.1   In both the L and the G versions of the *fourth* argument, Frege provides summary expositions of the route that leads to the sentences above. The problematic sentences come right after the corresponding definitions:

> (L10a) In each case in hand it would always come back to the question (L10b) whether it is true that *A* has such-and-such properties, or stands in such-and-such a relation to such-and-such a thing.

(G10a) And in application to any particular case it would always depend on

(G10b) whether it were true that the characteristics were present.

The individual steps leading to these sentences are not presented in the fourth argument. Nonetheless, this conciseness of exposition does not diminish the validity of the argument, as these steps are presented in detail in the preceding third argument and can be easily transferred to the fourth one.

7.4.2 Let us now compare the two versions of the third argument in L and G, respectively, with one additional step included in each case. Table 7.2 below offers a clear view on both versions.

Leaving the differences between the definitions aside, the steps that Frege uses in both arguments can be made explicit as follows:

I. We establish a definition of the form 'x is true:= x is P.'
II. We ask with regard to a given idea I_0,

whether I_0 is true.

III. We apply the definition to this particular case and substitute the definiens for the definiendum. Thus, we ask

whether I_0 is P.

IV. We apply (F*) to this interrogative sentence and ask

whether it is true that I_0 is P.

Frege does not walk through each of these steps one by one. Rather, in each of the arguments he includes a double step: in L this is steps II/III (see (L4*)), while in G it is steps III/IV (see (G4*)). Nonetheless, from a logical point of view all four steps are present in both arguments, which is why the one is just as valid as the other: combining two logical steps into one does not diminish the validity of a proof.

The substitution of the definiens for the definiendum in step III is unproblematic. Principle (F*), the omnipresence of truth, is explicitly endorsed by Frege, who asserts and uses the principle as a presupposition in both L and G (see Chapter 9). Therefore, under Frege's presupposition (F*), steps (L4) – (L5) and (G4) – (G5) are *obviously* justified.

Table 7.2   The third argument in 'Logik' and 'Der Gedanke'

(L4a) [...] in order to apply this definition we should have to decide in a particular case	(G4a) And what would we then have to do so as to decide
[(L4*) whether an idea is true in accordance with this definition, that is]	(G4b) whether something were true?
(L4b) whether an idea did correspond to reality,	[(G4*) that is, whether an idea and something real correspond in the specified respect.]
(L5a) or in other words:	(G5a) We should have to inquire
(L5b) whether it is true that the idea corresponds to reality.	(G5b) whether it were true that, for example, an idea and something real correspond *in the specified respect.*
(L6) *Thus we should have to presuppose the very thing that is defined.*	(G6) *And with that we should be confronted again by a question of the same kind, and the game could start all over.*

7.4.3   We can reconstruct the two versions of the fourth argument in L and G, respectively, in exact analogy to the third argument. In all four cases (both versions of the third and both versions of the fourth argument), we have essentially the same series of steps, which is justified under the presupposition of (F*). This appears obvious.

We still need to understand, however, why Frege does not *evaluate* the results in these four cases in the same way. Again, at the end of each series of steps we have one of the following sentences:

(L5b) whether it is true that the idea corresponds to reality.

(L10b) whether it is true that A has such-and-such properties, or stands in such-and-such a relation to such-and-such a thing.

(G5b) whether it is true that, for example, an idea and something real correspond in the specified respect.

(G10b) whether it were true that the characteristics were present.

Since definition (L2) in L is a special case of (L9), (L5b) has to reveal a circle if (L10b) does. This is obvious, too. But we are left with two remaining unanswered questions :

(1) Why does Frege abandon the circle objection with regard to (G5b)?

(2) Why does Frege uphold the circle objection with regard to (G10b)?

7.5 The error in the circle objection in 'Logik'

7.5.0 The circle objection is most explicit in the third argument in L:

(L2) If, for example, we wished to say: 'An idea is true if it corresponds to reality',

(L3) nothing would have been achieved, since

(L4a) in order to apply this definition we should have to decide in a particular case

(L4b) whether an idea did correspond to reality,

(L5a) or in other words:

(L5b) whether it is true that the idea corresponds to reality.

(L6) Thus we should have to presuppose the very thing that is defined.

Now, how does (L5b) show that *the very thing that is defined should have to be presupposed?* The thing that is defined in (L2) is the concept 'true'. If we apply the definition and substitute the definiendum (= the thing defined) by the definiens we obtain question (L4b), which is identical to question (L5b); both *questions* are identical because both interrogative *sentences* express the same content 'in other words' (see (L5a)). The second part of (L5b) ('the idea corresponds to reality') contains the definiens, and the first part ('whether it is true that') the thing defined (= the definiendum). Thus, in this sense the definiendum is 'presupposed' in the first part of the sentence.

Even if we grant that the defined concept 'true' appears in the first part of (L5b), it is still disputable whether this means that the definiendum really is *presupposed*, thus generating a circle. But I will show in Chapter 8 that Frege's outline of this proof can indeed be reconstructed as a valid circle argument for (G10b), namely, by adding another step in which the question is divided up into two subquestions.

At this point, we are not yet concerned with the validity of the argument but rather with the correct interpretation of Frege's text: in my view, (L6) means that *with the word 'true' the thing defined is presupposed* in the first part of (L5b). The application of this definition appears to work fine in (L4b) and in the second part of (L5b), but in the first part of (L5b) the thing defined is presupposed; this is how the circle arises.

7.5.1 Now if this is Frege's circle objection, why does he regard it as invalid in 'Der Gedanke'? Let us try and link (L6) to (G5b) as follows:

> (G5b) whether it were true that, for example, an idea and something real, correspond in the specified respect.
>
> (L6) Thus we should have to presuppose the very thing that is defined.

Here, (L6) would assert that the thing defined was presupposed in the first part of (G5b) ('whether it were true that'). If the circle objection is valid with regard to (L5b) then it must also be valid with regard to (G5b), for here, too, the definiens has been substituted for the definiendum in the second part of the sentence and the first part of the sentence contains the defined concept 'true' as a presupposition. The differences between the definitions in L and G versions of the third argument are so minor that they cannot have any impact on the efficacy of the circle objection. This makes the question of why Frege gives up the circle objection in the G version of the third argument even more puzzling.

Under scrutiny, it becomes obvious that Frege clearly distinguishes between two concepts of truth in the third and fourth paragraphs of the G text: I-truth as the improper truth, and S-truth as the proper, primary truth. The 'attempt to explain truth as correspondence [of ideas with something real]' comprises three definitions, each of which builds upon its predecessor, as well as their respective critiques in the first three arguments. In the fourth paragraph, I-truth is 'reduced to the truth of sentences'; such a reduction of one concept of truth to another would not be possible here unless *two different* concepts of truth were being discussed.

Now, it is the difference between these concepts of truth that sabotages the circle objection. As (G2) and the second part of (G5b) show, the thing that is defined here is that an *idea* is true if the idea corresponds to something real in a specified respect. Thus, the thing defined is *I-truth* as correspondence. But the first part of (G5b), that is, the expression 'whether it is true that', reveals not the previously defined I-truth but rather its competitor, *S-truth*. Now since these are two different concepts of truth, the expression 'whether it is true that' in (G5b) does not, as (L6) claims, presuppose 'the very thing that is defined'. Put more explicitly, (G5b) says this:

> (G5b) whether it were true [S-truth]

that, for example, an idea and something real correspond in the specified respect. [I-truth]

Thus, here in the second part the definiens of I-truth has been inserted, but the first part of the sentence does not contain the concept of I-truth but that of S-truth. Hence, (L6) would be an invalid objection if it were applied to (G5b).

7.5.2   The circle objection in (L6) is just as inappropriate with respect to (L2) – (L5b) if we draw a distinction between I-truth and S-truth in 'Logik' in the same strict manner in which Frege does in 'Der Gedanke'. In

(L2) [...] An idea is true if it corresponds to reality [= the idea is I-true]

Frege defines I-truth, while in

(L5b) [...] whether it is [S-]true

that the idea corresponds to reality [= the idea is I-true]

the first part of the sentence ('whether it is [S-]true') contains the concept of S-truth.

Presumably, in 'Logik' Frege may not have paid much attention yet to this difference in (L5b) and hence thought that the same concept of truth occurred both in the first and the second part of the sentence:

(Lb5) [...] whether it is true [= is true]

that the idea corresponds to reality [= is true].

But if we draw a strict line between I-truth and S-truth, as Frege does in his criticism of psychologism in 'Der Gedanke', then we can no longer maintain the circle objection against (L2) – (L5b). By replacing the sentence he uses in 'Logik'

(L6) Thus we should have to presuppose the very thing that is defined

with the corresponding one he uses in 'Der Gedanke'

(G6) And with that we should be confronted again by a question of the same kind, and the game could start all over he seems to be correcting a mistake that he made at the beginning of 'Logik'.

7.5.3   Though Frege states already in 'Logik' (the statement comes 2.5 densely printed pages after his proof of indefinability): 'Thoughts are fundamentally different from ideas (in the psychological sense)' (Frege, 1897, p. 142 (131)), he here distinguishes thoughts from *ideas in the psychological sense*. Could it be that he defined truth with respect to *non*-psychological ideas at the beginning of 'Logik'? Could it be that in

   (L2) [...] an idea is true if it corresponds to reality

he means by 'idea' any suitable truth bearer that is distinct from ideas in the psychological sense? In this case the circle objection in (L6) against (L2) – (L5b) could perhaps be defended. For in this case the first and second clause of

   (L5b) whether it is true that the idea corresponds to reality

could indeed refer to 'the same' truth. According to this interpretation, Frege does not commit an error in 'Logik', but his use of terminology becomes more differentiated in the course of the 20 years after 'Logik'. In any case, Frege can no longer uphold the circle objection in (L6) against (G5b) (or (L5b)) in 'Der Gedanke' simply because his differentiation between I-truth and S-truth renders (L6) invalid.

7.6   The transition from 'Logik' to 'Der Gedanke'

7.6.0   If indeed the *difference* of the truth concepts in the first and second parts of (G5b) (or (L5b)) renders the circle objection invalid, and this is noticed by Frege, then he cannot simply reiterate the third objection of 'Logik' in 'Der Gedanke' without making any changes. Two complementary considerations suggest themselves with regard to these changes.

First, the *difference* of the truth concepts in (G5b) can be used to argue *against* I-truth, since (G5b) does show the dependence of I-truth on S-truth, and this dependence disqualifies I-truth as a candidate for scientific truth. It is in this sense that I read the G version of the third argument. The difference between the two truth concepts occurring in (G5b) (or (L5b)) is, on one hand, the reason why the circle objection is invalid and Frege abandons it; the difference is, on the other hand, the point at which the dependence of I-truth on S-truth becomes obvious. Thus, the difference between these two concepts of truth, which is the centerpiece of my interpretation of the third argument, explains why Frege abandons the circle objection in the G version of the argument.[6]

Secondly, if *one and the same* truth concept indeed did occur in the first and second parts of

(G10b) whether it were true that the characteristics were present,

then the circle objection intended in L would be valid again in the fourth argument in G. If the *difference* between the two concepts of truth that occur in the first and second parts of (G5b), respectively, is Frege's reason for abandoning the circle objection, then the *identity* of the truth concepts in the first and second parts of (G10b) should be a reason to uphold the circle objection in the G version of the fourth argument, as Frege indeed does.

7.6.1 If these two complementary considerations are correct with regard to Frege's texts, we can represent the relation between L and G as follows. Given truth concepts $truth_1$ and $truth_2$, the question of

(*) whether it is $true_1$ that truth bearer x is $true_2$

shows that the applicability of $truth_2$ depends on that of $truth_1$. Now, we can distinguish two cases:

(A) If $truth_1 = truth_2$ then (*) shows the dependence of $truth_2$ on itself, hence a circle.

(B) If $truth_1 \neq truth_2$ then (*) shows the dependence of $truth_2$ on another truth, namely $truth_1$.

In the 'Logik' version of the third and fourth arguments Frege does not sufficiently distinguish between I-truth and S-truth, which is why he raises a circle objection in both arguments along the lines of (A). In 'Der Gedanke', by contrast, we have an occurrence of (B) in the third argument and an occurrence of (A) in the fourth *if* in the latter context S-truth = $truth_1$ = $truth_2$.

But does the fourth argument in G indeed deal exclusively with S-truth? Does S-truth also occur in the second part of (G10b)? In what follows I shall show that the fourth argument in 'Der Gedanke' can be indeed understood along these lines.

7.6.2   In 'Der Gedanke' the connection between the two arguments is established as follows:

> (G7) So this attempt to explain truth as correspondence [of ideas to 'something real'] breaks down.
>
> (G8) But likewise any other attempt to define truth also breaks down.

Truth as correspondence is I-truth, whose three components Frege has discussed in the first three objections. This attempt, which breaks down according to (G7), concerns I-truth. Since Frege does not even mention, let alone consider, any other attempts to define I-truth, I read the expression 'any *other* attempt' in (G8) as referring to the only other type of truth Frege discusses, that is, S-truth. Therefore, we should assume that *any other attempt to define truth* consists in a definition of the truth of *sentences/thoughts* as scientific truth.

In short: the G version of the fourth argument can be understood as rejecting any attempt to define *S-truth*.

7.6.3   Now, if we admit only sentences/thoughts as truth bearers A, the circle objection from 'Logik' still holds:

> (L9) 'A is true if and only if it has such-and-such properties or stands in such-and-such a relation to such-and-such a thing'.
>
> (L10a) In each case in hand it would always come back to the question
>
> (L10b) whether it is true [= S-true]
>
> that A has such-and-such properties, or stands in such-and-such a relation to such-and-such a thing [= that A is S-true].

If A is a sentence or thought then (L9) constitutes a definition of S-truth. The second part of (L10b) contains the definiens and the first part of (L10b) makes it clear that the very concept of S-truth that is defined is presupposed in application of this definition:

> (L6) Thus we should have to presuppose the very thing that is defined [namely S-truth].

The circle objection (L6) is valid in this case because here S-truth is the thing previously defined in (L9). Hence, the fourth argument could be essentially transplanted from 'Logik' into the later essay 'Der Gedanke', provided that the 'thing that is defined' in (L9) is exclusively S-truth.

7.6.4 If the fourth objection in 'Der Gedanke' is aimed at definitions of *S-truth*, the first part and the second part of (G10b) refer to the same truth concept:

> (G9) For in a definition [of S-truth] certain characteristics would have to be specified.
>
> (G10a) And in application to any particular case [a specific sentence/thought] it would always depend on
>
> (G10b) whether it were true [= S-true] that the characteristics were present [in the sentence/thought] [= that the sentence/thought were S-true].
>
> (G11) So we should be going round in a circle.

This is essentially the same argument as in 'Logik'.

It remains to be examined whether the *characteristics* here are supposed to be (i) both absolute and relational properties, or (ii) only absolute properties. In case (i) the circle objection would be directed also against any relational definition of S-truth (such as S-truth as correspondence). Since, however, it already follows from Frege's first argument against I-truth in 'Der Gedanke' that *any* relational definition of truth necessarily misses the concept of absolute truth as used in science, I consider it more likely that Frege here only considers definitions by means of absolute properties. Moreover, in no place in the third paragraph does Frege relate the concept of correspondence to that of S-truth; rather, he identifies truth as correspondence with I-truth. This also supports the conclusion that *any other attempt* (see (G8)) at a definition of truth as critiqued in the G version of the fourth argument consists in defining S-truth as absolute truth and hence by means of absolute properties.

Nonetheless, I shall also show in Chapter 8 that the circle objection succeeds even for relational definitions of S-truth.

7.7 The respective contexts in 'Logik' and in 'Der Gedanke'

7.7.0 Frege makes a much stricter distinction between I-truth and S-truth in 'Der Gedanke' than in 'Logik' so that he has to give up the

circle objection developed in the third argument in 'Logik'. This, however, cannot be concluded solely on the basis of the text fragments G (=
(G1) – (G12)) and L (= (L1) – (L12)); we must also thoroughly consider the
contexts in which these fragments occur.

If we compare the initial pages of 'Logik' with those of 'Der Gedanke',
we notice far-reaching similarities in the theses but also significant
modifications in the argumentation, in particular, a much stricter distinction between I-truth and S-truth. In what follows I should like to
illustrate this point by means of a few examples.

7.7.1   In the first paragraph of 'Logik' Frege emphasizes how important
the word 'true' is for logic, and in the second paragraph he characterizes logic as 'the science of the most general laws of truth' (1897, p. 139
(128)). This corresponds to the first paragraph in 'Der Gedanke', except
that the first and second 'Logik' paragraphs do not contain any criticism of psychologism. Frege writes in the second paragraph of 'Logik':

> We must assume that the rules for our thinking and for our holding
> something to be true are prescribed by the laws of truth. The former
> are given along with the latter. Consequently we can also say: logic
> is the science of the most general laws of truth. (Frege, 1897, p. 139
> (128))

Obviously, here the rules for holding something to be true are unprob-
lematically placed alongside the logical laws of truth, and Frege does
not see the need at this point in 'Logik' to launch an attack against
psychologism. The situation is very different in the first paragraph of
'Der Gedanke', where he distinguishes very strictly between logic and
psychology:

> From the laws of truth there follow prescriptions about asserting,
> thinking, judging, inferring. And we may very well speak of laws of
> thought in this way too. But there is at once a danger here of con-
> fusing different things. People may very well interpret the expres-
> sion 'law of thought' by analogy with 'law of nature' and then have
> in mind general features of thinking as a mental occurrence. A law
> of thought in this sense would be a psychological law. And so they
> might come to believe that logic deals with the mental process of
> thinking and with the psychological laws in accordance with which
> this takes place. That would be misunderstanding the task of logic,
> for truth has not been given here its proper place. [...] In order to

avoid any misunderstanding and prevent the blurring of the boundary between psychology and logic, I assign to logic the task of discovering the laws of truth, not the laws of taking things to be true or of
thinking. (Frege, 1918a, pp. 58–9)

Even more pointedly, Frege writes in a later paragraph in 'Der
Gedanke':

> Not everything is an idea. [...] the scientist will surely not acknowl-
> edge something to be the firm foundation of science if it depends on
> men's varying states of consciousness [as ideas do]. [...]
>
> Not everything is an idea. Otherwise psychology would contain all
> the sciences within it, or at least it would be the supreme judge over
> all the sciences. Otherwise psychology would rule even over logic and
> mathematics. But nothing would be a greater misunderstanding of
> mathematics than making it subordinate to psychology. [ibid., p. 74]

In this context Frege repeats three (!) times the sentence 'Not everything
is an idea' (ibid., pp. 73–4). The fundamental concept of psychologism
is the concept of an idea. If ideas were admitted as truth bearers, the
boundary between logic and psychology would be in danger of becoming blurred.

7.7.2 Frege does briefly mention our common talk of *true ideas* in
'Logik' too; but he does not here develop it into a counter-position to be
attacked, as he does in 'Der Gedanke'; instead, he immediately reduces
the truth of ideas to that of thoughts:

> [(A)] Of course we speak of true ideas [...] as well. By an idea we under-
> stand a picture that is called up by the imagination: unlike a percep-
> tion it does not consist of present impressions, but of the reactivated
> traces of past impressions or actions. Like any other [!] picture, an
> idea is not true in itself, but only in relation to something to which
> it is meant to correspond.[7] [(B)] If it is said that a picture is meant
> to represent Cologne Cathedral, fair enough; it can then be asked
> whether this intention is realized; if there is no reference [Hinblick]
> to an intention to depict something, there can be no question of the
> truth of a picture. [(C)] It can be seen from this that the predicate *true*
> is not really conferred on the idea itself, but on the thought that the
> idea depicts a certain object. And this thought is not an idea, nor is

it made up of ideas in any way. Thoughts are fundamentally differ-ent from ideas (in the psychological sense). The idea of a red rose is something different from the thought that this rose is red. Associate ideas or run them together as we may, we shall still finish up with an idea and never with something that could be true. (Frege, 1897, p. 142 (131))

Now, all of what Frege says about ideas and thoughts in this quoted passage conforms with what he says in 'Der Gedanke'. However, in the latter text he says a lot *more* on the topic of I-truth. In 'Logik' Frege introduces the concept of an idea only *in passing*: '*Of course, we speak of true ideas as well*' (see (A), my emphasis). In 'Der Gedanke', by con-trast, I-truth is introduced as a *conception of truth in its own right*: (β) 'It might be supposed from this that truth consists in a correspondence of a picture to what it depicts' (iv). This conception of I-truth is then discussed thoroughly in three consecutive definition attempts – which are indeed adequate *as* definitions of I-truth! – and three corresponding objections. At the end of the third objection Frege writes: (ββ) 'So this attempt to explain truth as correspondence breaks down' (vi).

  'Logik', by contrast, does not contain anything like a genuine attempt to discuss I-truth as a candidate for scientific truth. Though the above passage contains an indication of the *relational* character of the truth of ideas – 'Like any other picture, an idea is not true in itself, but only in relation to something to which it is meant to correspond' (see (A)) – the expression 'true ideas' is immediately rejected and reduced to the truth of thoughts: 'It can be seen from this that the predicate *true* is not really conferred on the idea itself, but on the thought that the idea depicts a certain object' (see (C)).

  This is essentially the same reduction of I-truth to S-truth that later occurs in 'Der Gedanke', but here in 'Logik' it is not *explicitly* designated as such. Since in 'Logik' I-truth is not even discussed as a conception of truth in its own right, the question of the reducibility of one truth con-cept to another does not arise. Furthermore, Frege speaks of 'alterations of sense' only in 'Der Gedanke', not in 'Logik'. In 'Der Gedanke' I-truth is placed opposite S-truth as a conception of truth in its own right and thoroughly discussed. Since Frege here starts out by taking I-truth seri-ously as an opposing view, he must later disqualify it as exemplifying an 'improper' use of the word 'true' and *explicitly* say that it can be 'reduced' to S-truth:

'My idea corresponds to Cologne Cathedral' is a sentence, and now it is a matter of the truth of this sentence. So what is *improperly* called the truth of pictures and ideas is *reduced* to the truth of sentences. (viii)

7.7.3   The examples cited in this section show that Frege makes a much stricter distinction between I-truth and S-truth in 'Der Gedanke' than he does in 'Logik'. And it is precisely because of this much clearer distinction between the two concepts of truth that he has to abandon the circle objection in the third argument.

7.8   Concluding comparison of L and G

The parallel passage in 'Logik' (L1) – (L12) constitutes a self-contained argumentation with only one argumentative goal: to prove the indefinability of truth. The fourth argument alone would suffice to achieve this argumentative goal, but Frege precedes it with the third argument as a special case. Both arguments are, being circle objections, *formal* arguments.

By comparison, the parallel passage in 'Der Gedanke' (G2) – (G12) is taken out of the context of the third and fourth paragraphs, and without that context the third argument can hardly be understood properly. The third argument in 'Logik' is replaced in 'Der Gedanke' by *three* closely interconnected arguments. In these three arguments the three components of I-truth are presented and critically discussed with respect to their *content*. The argumentative goal of the first three objections in 'Der Gedanke' is to disqualify I-truth as a candidate for scientific truth, and thereby also ideas as truth bearers. The argumentative goal of the fourth argument in 'Der Gedanke' is to show that S-truth is indefinable. Since according to the first three content-based arguments scientific truth cannot be defined as I-truth, and since S-truth, which is identical to scientific truth, cannot be defined due to the formal circle objection, it follows from all four arguments in conjunction that scientific truth is indefinable.

In the third and fourth arguments in 'Der Gedanke', Frege clearly distinguishes between I-truth and S-truth. In 'Logik', by contrast, he draws no such clear distinction and certainly not as early as in the context of his argument for the indefinability thesis. This is why Frege raises the circle objection in the third argument in L, while abandoning it in G. Hence, Frege's distinction between I-truth and S-truth is of fundamental importance for any serious attempt at understanding the differences between the two parallel text passages.

Table 7.3 provides a summarized comparison of the two texts and their differences.

Table 7.3   Comparison of L and G

'Logik'	'Der Gedanke'
1 truth concept	2 truth concepts: I-truth and S-truth, in competition for the status of (proper) truth
1 argumentative goal: (A) indefinability of truth	2 argumentative goals: (A) indefinability of truth (B) exclusion of I-truth as scientific truth and of ideas as truth bearers
1 formal argument for (A): circle objection in two variants (a) against truth as correspondence of ideas, (b) against all definitions of truth whatsoever. The formal third argument is a circle objection with regard to *one* (undifferentiated) concept of truth.	4 arguments for (A) or (B): 3 content-based arguments for (B) 1 formal argument for the indefinability of S-truth: circle objection (A) emerges as a result of all 4 arguments in conjunction. The formal third argument in 'Logik', the circle objection, turns out to be invalid due to the differentiation of *two* concepts of truth. It is replaced by three content-based arguments.
(L1) – (L12) is a self-contained argumentation in favor of (A) that can be understood independently of the remainder of the text.	(G2) – (G12) is taken out of its context and can be understood only in the context of the third and fourth paragraphs.

7.9   The treadmill

7.9.0   There may be yet another parallel passage in 'Logik', the treadmill argument. Künne and other scholars read this argument as a paraphrase of the third and fourth arguments in 'Der Gedanke' (Künne, 2003, pp. 129, 132). However, the treadmill argument has a very different argumentative goal. I shall begin by quoting the argument in its context, and then provide a paraphrase, and finally compare it to the third and fourth arguments in 'Der Gedanke'.

Frege develops the argument in the following manner and context:[8]

> [I.] *A thought does not belong specially to the person who thinks it, as does an idea to the person who has it: everyone who grasps it encounters it in the same way, as the same thought.* Otherwise two people would never attach the same thought to the same sentence, but each would have

his own thought; and if, say, one man put 2 · 2 = 4 forward as true whilst another denied it, there would be no contradiction, because what was asserted by one would be different from what was rejected by the other. It would be quite impossible for the assertions of different people to contradict one another, for a contradiction occurs only when it is the very *same* thought that one person is asserting to be true and another to be false. So a dispute about the truth of something would be futile. There would simply be *no common ground to fight on*; [...]

[II.] If we wanted to regard *a thought as something psychological, as a structure of ideas*, without, however, adopting a *wholly* subjective standpoint [as the one above], we should have to explain the assertion that 2 + 3 = 5 on something like the following lines 'It has been observed that with many people certain ideas form themselves in association with the sentence "2 + 3 = 5". We call a formation of this kind the sense of the sentence "2 + 3 = 5". So far as we have observed hitherto these formations are always true; we may therefore make the provisional statement "Going by the observations made hitherto, the sense of the sentence '2 + 3 = 5' is true".'

But it is obvious that this explanation would not work at all. And *it would leave us where we were*, for the sense of the sentence 'It has been observed that with many people certain ideas form themselves etc.' would of course be *a formation of ideas* too and the whole thing would begin over again.[...]

[III.] *If a thought [G_A], like an idea, were something private and mental*, then the truth of a thought could surely only consist in a relation to something [A] that was not private or mental. So if we wanted to know

[(T1)] whether a *thought* [G_A] was true,

we should have to ask

[(T2)] whether the *relation* in question [between G_A and A] obtained and thus

[(T3)] whether the *thought* [G_{A*}] that this relation [between G_A and A] obtained was true.

And so we should be in the position of a man on a *treadmill* who makes a step forwards and upwards, but the step he treads on keeps giving away and *he falls back to where he was before*.

7.9.1 In these three passages, Frege argues for the thesis that thoughts exist independently of the thinker. That thesis is presented right at the start of the first paragraph:

[Thesis] A thought does *not* belong specifically to the person who thinks it, as does an idea to the person who has it, *but stands opposite everyone who grasps it in the same way, as the same thought.*

In each of the three paragraphs the corresponding counter-thesis of psychologism is refuted. The details of the argumentation are as follows.

The counterthesis can be made explicit in the first paragraph simply by deleting the word 'not':

[Counter-thesis] The thought does [...] belong specifically to the person who thinks it, as does an idea to the person who has it.

But if the thought belongs to the inner world of the person who thinks it, just as ideas do, then people could not contradict one another since they would not succeed in assigning truth or falsity to the *same* thought. While I hold *my* thought$_{U.P.}$ that 2 + 3=5 to be true (or false), another person would hold *his* thought$_{J.D.}$ that 2 + 3=5 to be false (or true). And this is an absurdity: 'It would be quite impossible for the assertions of different people to contradict one another; [...] There would simply be no common ground to fight on'.

Then in the two subsequent paragraphs, Frege considers two possible attempts at defending the counter-thesis, both of which fail. In paragraph II the counter-thesis is modified as follows:

If we wanted to regard a thought as something psychological, as a structure of ideas, without, however, adopting a *wholly* subjective standpoint [as the one above], [...]

In this defense a thought is still regarded as an idea, just as before, but the *wholly* subjective standpoint is supposed to be reduced to an *intersubjective* standpoint. According to this standpoint, truth or falsity are not assigned to only one's own inner formations of ideas but to *all* other formations of ideas as well that are associated with the same sentence. Thus, by reading the sentence 'It is true that 2 + 3=5' as meaning 'All inner structures of ideas that are associated with the sentence "2 + 3=5" in me or other people are true', I seemingly leave my inner world and connect to the inner worlds of other people, which provides me with some kind of common ground to fight on and the possibility that several people can contradict one another. But this is just appearance, for the sense of the sentence 'All inner structures of ideas that are associated with the sentence "2 + 3=5" in me or other people are true' would be nothing but, as Frege puts it, an inner 'formation of ideas too' and 'the whole thing would begin over again.' In short: taking recourse to intersubjectivity does not solve the problem of how to escape the confines of one's own inner world of formations of ideas.

 The counter-thesis is taken up again in paragraph III: 'If a thought [G_A], like an idea, were something private and mental, [...]' but the *wholly subjective* standpoint is now tentatively modified into an *object-based* standpoint. This relation to an object, however, can only be guaranteed by way of a relational concept of truth – or as Frege puts it: 'then its truth could surely only consist in a relation to something [A] that was not private or mental.' If *my* thought $G_{A(U.P.)}$ and another person's thought $G_{A(J.D.)}$ referred to the *same* external object A, it would then be possible for us to contradict one another *even though* both of us would assign our truth values to our own private inner thoughts. I may then ask 'whether my thought G_A stands in this *relation* to A'.

By asking this question I try to escape from my inner world, to which G_A belongs, and to investigate the relation between my inner thought G_A and the external object A. But in fact this question does not succeed in leading me beyond the confines of my inner world, for it is not different from the question of 'whether the *thought* G_A* that this relation obtains [between G_A and A] is true'. The point here is that A is not given to me directly as something external in such a way that I could simply leave my inner world by contemplating A. Rather, A is given to me only as a component of my *thought* G_A*, and since G_A*, like all thoughts according to the above counter-thesis, is once again 'like an idea, [...] something private and mental', my attempted escape from my inner world must fail. My attempt to walk out of my inner world on some kind of plank connecting my inner thought G_A with the entity A yonder outside in the external world is doomed to fail because the plank keeps giving way just like in a treadmill. In short: A *relation to an external object* does not solve the problem of how to escape the confines of one's own inner world of formations of ideas either.[9]

7.9.2 According to the definition proposed here, the truth of a thought seems to consist in a certain relation to an external object:

> [The] truth [of an inner thought G_A] could surely only consist in a relation to something [A] that was not private or mental [see above].

Frege considers this relational definition of truth of thoughts *only under the presupposition* that 'a thought, like an idea, were something private and mental'. Thus, there is no evidence here that Frege discusses a relational concept of truth for *thoughts* as he does for ideas. Though he does this for ideas, he does it for thoughts only under the assumption that they are ideas or something private and mental just like ideas (see for this also Chapter 12).

This also makes it clear why Frege takes it for granted that the truth of ideas – assuming them to be truth bearers – could only consist in their correspondence to something external. The reason is just that ideas are genuinely private, so that it would not be possible for different people to contradict one another without such a relational concept of truth – there would not be any common ground to fight on.

Conversely, it also becomes clear here why Frege does not apply the correspondence theory to thoughts: thoughts do not belong to the inner world, they are already out there on the common battle ground ready to be 'grasped' by any human being. Thus, we do not need correspondence in the case of thoughts in order to explain how different human beings can contradict one another.

7.9.3   The structure of the treadmill argument is similar to that of the third and fourth arguments in 'Der Gedanke'. Here too, as in 'Der Gedanke' and the parallel passage from 'Logik' discussed earlier, the argument is based on certain propositional questions to which (F*) and a definition of truth are applied. The propositional questions are as follows:

(T1) whether a thought $[G_A]$ was true,

(T2) whether the relation in question [between G_A and A] obtained,

(T3) whether the thought $[G_A{}^*]$ that this relation [between G_A and A] obtained was true.

Analogously to the two parallel passages G and L, (T2) is derived from (T1) by replacing the definiendum by the definiens, and (T3) is derived by application of (F*) from (T2).

Frege ends the *reductio ad absurdum* he performs in these three passages with very similar sentences, namely, (G5b), (G10b) and (T3), respectively. The descriptions of the absurdity, however, differ. In the third argument the propositional question is:

(G5b) whether it were [S-]true that, for example, an idea and something real correspond in the specified respect [that the idea is I-true].

In (G5b) the second part contains the defined concept of I-truth and the first part reveals that I-truth cannot be applied without S-truth:

(G6) And with that we should be confronted again by a question of the same kind, and the game could start all over.

The fourth argument contains the propositional question:

(G10b) whether it were [S-]true that the characteristics were present [in the thought] [= that the thought were S-true].

The second part of (G10b) contains the defined concept of S-truth and the first part reveals that the definition of S-truth cannot be applied without a circle, or in Frege's own words:

(G11) So we should be going round in a circle.

The propositional question in the treadmill argument is this:

(T3) whether the thought $[G_A{}^*]$ that this relation [between G_A and A] obtained was true.

The attempt to leave our inner world behind by employing a relational concept of truth and trying to grasp A as an external object that, as such, would be accessible to other people fails; it turns out that A can be grasped only as a component of the *thought* $G_A{}^*$, which belongs to the inner world. This is why it is impossible for us to leave our 'inner world':

And so we should be in the position of a man on a treadmill who makes a step forwards and upwards, but the step he treads on keeps giving way and he falls back to where he was before.

In contrast to his procedure in the third and fourth arguments, Frege presents a *reductio* not against the definition of truth in question, but against a very different assumption that is presented as a counter-thesis at the start of paragraph III (cf. the counter-theses in paragraphs I and II):

If a thought $[G_A]$, like an idea, were something private and mental, [...]

The assumption of the indirect proof here is this: a *thought* is, like an idea, something private. This is why the first clause of the propositional question is also formulated slightly differently:

> whether *the thought* [G_A*] that [...] was true

In comparison to this, the four analogous sentences in L and G only have an 'it' in place of 'the thought':

> whether *it* were (is) true that...

If the argumentative goal were to show the failures of a definition of truth then 'it' would suffice. But if the argument is about how to reach a 'common ground to fight on' from one's own inner thought via its truth – conceived as relational to something external – then (T3)'s use of 'the thought' is preferable, since it makes it explicit that we will not get any further than to a thought, once again conceived as an inner structure of ideas.

The metaphor of the treadmill is anticipated already in the following passage in paragraph II:

> And *it would leave us where we were*, for the sense of the sentence: 'It has been observed that with many people certain ideas form them-selves etc.' would of course be a formation of ideas too and the whole thing would begin over again.

This is just like a treadmill, for a treadmill also leaves us where we were before.

7.9.4   Künne more or less identifies the three arguments with one another when he writes:

> [Frege] tries to show, in part (A) [= the third argument in 'Der Gedanke'] of his argument, that the predicament of an advocate of Object-based Correspondence is that of 'a man in a treadmill who makes a step forward and upwards, but the steps he treads on keep giving way and he falls back to where he was before'. [...] If his argu-ment were convincing, it would also refute all varieties of Fact-based Correspondence, for as Frege himself registers in part (B) [= the fourth argument in 'Der Gedanke'] of his argument, it is actually an objec-tion against *any* attempt at defining the concept of truth.[10]

However, I consider this way of identifying the three arguments to be misguided. The similarity between the three arguments essentially consists in the fact that they all presuppose the omnipresence of S-truth; aside from this, the arguments are quite different, particularly with respect to their argumentative goals. In one of them Frege shows that I-truth is dependent on some other concept of truth; in another he shows that S-truth is indefinable; in the last he shows that thoughts cannot be inner ideas. The mere fact that Frege makes use of the omnipresence of S-truth in an argument by itself indicates nothing about whether or not he constructs a circle, or whether he intends to refute a certain definition of truth or rather something else. In order to decide these questions we must sufficiently take into account the context in which these arguments are made.

8
The Fourth Argument: The Circle Objection

8.1 The text

(vii) [Conclusion of the circle argument:] But likewise, any other attempt to define truth also breaks down.

[The circle argument:] For in a definition certain characteristics [note the plural!] would have to be specified. And in application to any particular case it would always depend on whether it were true that the characteristics were present. So we should be going round in a circle.

[Final conclusion of all four objections:] So it is likely that the content of the word 'true' is sui generis and indefinable. [End of the excursion]

According to the received view that has dominated Frege scholarship since the publication of Dummett's first book on the subject, Frege's third and fourth objections are more or less the same; furthermore, they are generally described as 'Frege's regress argument against the correspondence theory' of truth.[1] In my view, however, the similarity between the two arguments merely consists in the fact that both presuppose the omnipresence of truth and thus have *one shared presupposition*. But this presupposition is the departure point for two otherwise completely different arguments.

The following points seem to me crucial for arriving at an accurate interpretation of the fourth objection (I shall address each of these issues in more detail later on): first, when Frege speaks of a definition here he means an *analytic* definition, providing a conceptual analysis of the concept of truth into *several characteristics* A, B, ... – hence the special significance of the plural in 'certain characteristics'. Second, Frege says that the circle emerges 'in application to any particular case'; therefore it could be possible that the circle does not emerge in the definition itself, but only in its *application*. Third, we should trust that Frege is aware of the difference between a regress and a circle; when he writes 'So we should be going round in a circle', then, he surely intends to refer to a *circle* and not a regress. Fourth, following my results in section 7.6, I shall assume that when Frege speaks of 'any *other* attempt to define truth', he is referring to any attempt to define *S-truth* as scientific truth.

The basic idea of Frege's circle objection can be easily understood if we keep in mind that the circle is said to emerge in an *application* of an *analytic* definition of truth. In the following I shall begin with an outline of the basic structure of the argument and follow up with a more detailed discussion of some disputed issues concerning this argument.

8.2 An analytic definition with two one-place predicates

8.2.0 In the previously discussed parallel passage from 'Logik', Frege writes, 'Truth is obviously something so primitive and simple that it is not possible to reduce it to anything still simpler.' A definition in which truth is reduced to another thing still *simpler* – or a definition in which certain *characteristics* (note the plural!) would have to be specified, as required in (vii) – would be what Frege terms an analytic definition. He distinguishes between a *constructive* and an *analytic* type of definition, a distinction which corresponds to that between a nominal definition and a conceptual analysis. In this connection he brings up the notion

> that it is by means of definitions that we perform logical analyses. In the development of science it can indeed happen that one has used a word, a sign, an expression, over a long period under the impression that its sense is simple until one succeeds in analysing it into simpler logical constituents. By means of such an analysis, we may hope to reduce the number of axioms; for it may not be possible to prove a truth containing a complex constituent so long as that constituent remains unanalysed; but it may be possible, given an analysis, to prove it from truths in which the elements of the analysis occur. [...]

We have therefore to distinguish *two quite different cases*:

(1) We construct a sense out of its constituents and introduce an entirely new sign to express this sense. This may be called a 'constructive definition', but we prefer to call it a 'definition' *tout court*.

(2) We have a simple sign with a long established use. We believe that we can give a logical analysis of its sense, obtaining a complex expression which in our opinion has the same sense. We can only allow something as a constituent of a complex expression if it has a sense we recognize. The sense of the complex expression must be yielded by the way in which it is put together. That it agrees with the sense of the long established simple sign is not a matter for arbitrary stipulation, but can only be recognized by an immediate insight. No doubt we speak of a definition in this case too. It might be called an 'analytic definition' to distinguish it from the first case. But it is better to eschew the word 'definition' altogether in this case, because what we should here like to call a definition is really to be regarded as an axiom. In this second case there remains no room for an arbitrary stipulation, because the simple sign already has a sense. (Frege: 1914, 226f (209f))

An *analytic* definition such as

x is a bachelor:= x is male and x is unmarried,

enables us to conduct an analysis of the concept 'bachelor' into simpler constituents. By contrast, a *constructive* definition such as

x is mu:= x is male and x is unmarried,

enables us to introduce an abbreviation, for the definition allows us to use the shorter expression 'mu' in place of the definiens. But the definition

Aubergine:= Eggplant

is not an analytic definition either, for the definiens 'eggplant' does not constitute a 'complex' expression of which we believe that it expresses the same sense as the word 'aubergine', which is why the meaning of 'aubergine' is not reduced to something simpler in this case. Nor is

'Aubergine:= Eggplant' a constructive definition, for 'aubergine' is not 'an entirely new sign'; in particular, 'aubergine' is not an abbreviation of 'eggplant'. Rather, a simple (non-complex) expression is here replaced by another simple (non-complex) expression.

8.2.1 Let us now consider an analytic definition in which truth is defined by means of *two characteristics*, that is, one that contains two one-place predicates that are connected by the sentential operator 'and':

x is true:= x is A and x is B.

Let us assume that this definition is non-circular; that is, neither of the two predicates 'x is A' and 'x is B' contains the predicate 'x is true' or any predicate synonymous with it. Then, when applying the analytic definition in a concrete case, we can *analyze* (or *decompose*) a complex initial question such as

(1) Is it true that it is raining?

into the two simpler subquestions:

(1a) Is it A that it is raining?

and

(1b) Is it B that it is raining?

This *decomposition* into subquestions is typical of the application of an analytic definition and represents a crucial step in this argument. If both subquestions can be answered with 'Yes', we are also entitled to answer question (1) with 'Yes'.

Since the definition is non-circular, neither of the *interrogative sentences* (1a) and (1b) contains the term 'true' (or another term synonymous with it). At this point, the definition *looks* successful: the complex question of the truth of a given thought can be analyzed into two simpler questions about whether the thought has the characteristics A and B, respectively. Since neither of the two subquestions contains the truth predicate, the aim of reducing truth 'to anything still simpler' (Frege, 1897, p. 140 (129)), modified translation) appears to have been successful.

Yet, though the definition itself is non-circular, a circle can be shown to emerge in its *application*. For according to Frege's presupposition

(F*) The interrogative sentences 'p?' and 'Is it true that p?' are synonymous, that is, they have the same Fregean sense (see section 6.4).

the interrogative sentences (1a) and (1b) are *synonymous* with the following extended interrogative sentences:

(2a) Is it true that it is A that it is raining?

and

(2b) Is it true that it is B that it is raining?

Although the *interrogative sentences* (1a) and (1b) are distinct from the interrogative sentences (2a) and (2b), they are synonymous with them; and this is why *questions* (1a) and (1b) are identical to questions (2a) and (2b), respectively. Hence, we cannot answer either of the subquestions (1a) and (1b) unless we already know what truth is, insofar as these subquestions are identical to the subquestions (2a) and (2b), whose formulations *explicitly* contain the term 'true'. Thus, even if our analytic definition of truth is not itself circular, a circularity emerges in its *application*, rendering the definition futile.

We could prove the existence of a circle even if the definition contained a different sentential operator from 'and'. In case of the disjunctive operator 'or' we would need to answer only *one* of the subquestions (1a)/(2a) or (1b)/(2b) in the affirmative in order to have a positive answer to the more complex question (1); but even with that one question we would be able to discover the application circle. Similar conditions apply to other sentential operators. In the following I shall focus exclusively on the conjunctive operator 'and', with the exception of sections 8.4 and 8.5.

8.2.2   In a circular definition such as

x is true:= x is A and x is true.

we immediately observe an application circle. In this case we do not need to bring (F*) to bear on the subquestions – which in this case are

(1a*) Is it A that it is raining?

and

(1b*) Is it true that it is raining?

– since question (1b*) already contains the concept of truth explicitly.

In short: every circular definition leads to an application circle, but not every definition leading to an application circle is a circular definition.

In a nutshell, the difference between the two types of circle is this. If we need the defined concept in order to apply an analytic definition to a particular case then we have a *circular application*. If, on the other hand, the defined concept is contained, either explicitly or implicitly, in the *definiens* of an analytic definition (so that it can be disclosed as a component of the definiens in the process of analysis), then we have a *circular definition* (see Chapter 9).

8.2.3   I would like to further clarify the notion of an application circle by considering an example of a definition whose application is non-circular. Let us take the following familiar example:

x is a bachelor:= x is a man and x is unmarried.

This analytic definition can be applied entirely without circularity. To apply it, we replace the question
    Is Peter a bachelor?
with the two subquestions
    Is Peter a man?
and
    Is Peter unmarried?
each of which we can answer without knowing what a bachelor is. We can see this already just by looking at the first subquestion 'Is Peter a man?', which is not at all identical to the (nonsensical) question
    Is it a bachelor that Peter is a man?
In contrast, the analogous questions derived from the definition of truth under consideration – 'Is it A that it is raining?' and 'Is it true that it is A that it is raining?' – are identical.

8.2.4   Up to this point my interpretation of the circle objection fully reflects Frege's own text, as the following paraphrase shows:

> (vii*) But likewise, any other attempt to define truth also breaks down. For in a definition certain characteristics [A and B] would have to be specified. And in application to any particular case ['Is it A that p?'; 'Is it B that p?'] it would always depend on whether it were true that the characteristics were present ['Is it true that it is A that p?'; 'Is it true that it is B that p?']. So we should be going round in a circle.

Obviously, the fourth argument is very different from the third one.

8.3   The inapplicability of the circle objection to a non-analytic definition

8.3.0   The above circle demonstration, which concerns the application of an analytic definition, proceeds by way of the following steps:

(1) The concept of truth is *decomposed* by the definition into characteristics A and B.
(2) We apply the definition to the question: 'Is it true that p?' by substituting the definiens for the definiendum; we then obtain the synonymous question: 'Is it A and B that p?'
(3) We then *decompose* the complex question 'Is it A and B that p?' (= 'Is it true that p?') into the simpler subquestions 'Is it A that p?' and 'Is it B that p?', neither of which contains a truth predicate.
(4) Finally, we show by means of (F*) that, nonetheless, the two subquestions do contain the truth concept implicitly (since both interrogative sentences are synonymous with interrogative sentences containing a truth predicate explicitly).

The application of our analytic definition takes place in the second and third steps. In the following I shall discuss in more detail the significance of the third step, which consists in a decomposition of the complex question into subquestions, for the process of demonstrating the presence of the circle. There are many reconstructions of Frege's circle objection in which this decomposition is missing, and these reconstructions do not reveal a valid circle argument. In some of them even the first step is missing, that is, they do not even start out with an *analytic* definition (which, of course, also excludes the very possibility of a decomposition into subquestions).

8.3.1 I wish to show, first, that in applying a *non-analytic* definition such as

x is true:= x is T.

a circle cannot be revealed in principle. Suppose we apply this definition (which does not offer us any 'characteristics' in the plural, as Frege requires) to the question

(1) Is it true that it is raining?

Then, by substitution of the definiens for the definiendum (see step 2), we obtain the equally complex question

(3) Is it T that it is raining?

This step is entirely unproblematic and does not involve any flaws. However, the third step, the decomposition of the complex question (1)/(3) into subquestions is not possible here, since the definition is non-analytic. Thus, we do not get beyond interrogative sentence (3) in our application of the definition, and (3) by itself contains a truth predicate just as obviously and *explicitly* as the original interrogative sentence (1). As a result, there is not even an appearance, as there is in the earlier example above, of the complex question (1)'s having been replaced by several simpler questions that do *not* contain the truth concept (the sense of 'true'). Since 'T' has the same sense as 'true' we will not succeed in getting rid of this very *sense* by replacing the *word* 'true' by the synonymous word 'T' or, in other words, by moving from (1) to (3).

It is obvious that in the above example there is a truth predicate in both interrogative sentences (1) and (3), that is an expression that has the sense of 'true'. In contrast to this, in the transition from (1) to the subquestions

(1a) Is it A that it is raining?

and

(1b) Is it B that it is raining?

it *appears* as if we obtained *questions* that do *not* contain the truth concept, since the two *interrogative sentences* (1a) and (1b) do not contain any truth predicate. But the application of

(F*) The interrogative sentences 'p?' and 'Is it true that p?' are synonymous

to (1a) and (1b) reveals that this is only an appearance and that we did not, after all, get rid of the truth *concept* in questions (1a) and (1b), even though the corresponding interrogative *sentences* do not contain a truth *predicate*. This, is, in fact, the pattern for discovering non-obvious circularities. The circularity is not obvious precisely because it *appears* we can do without the definiendum (here, 'truth') – but is nevertheless there because in *reality* we cannot.

As contrasted with an analytic definition, in a non-analytic definition of truth such as 'x is true:= x is T' it is obvious that we cannot do without the defined *concept* (*true/T*) – we do not need to employ (F*) in order to see this.

8.3.2 Now if in addition we apply (F*) to

(3) Is it T that it is raining?

Then we obtain

(4) Is it true that it is T that it is raining?

Since 'true' and 'T' are synonymous expressions, (4) is synonymous with

(4a) Is it T that it is T that it is raining?

Now it would be absurd to diagnose a *circle* in the transitions from (1) to (3) and then to (4) or (4a), since *each* of these interrogative sentences contains at least one truth predicate and no attempt is made to formulate interrogative sentences without a truth predicate. And if the transition from (1) to (3) does not constitute a circle – as it surely does not! – then why should the transition from (3) to (4) or (4a)? The fact that (3) contains *one* truth predicate and that (4) contains *two* truth predicates and (4a) *two* tokens of the same truth predicate does not establish that the transition from (3) to (4) or (4a) is a circle. Or else we would have to conclude that the transition from 'p' to 'p or p' is also a circle because 'p' occurs twice in the latter sentence.

If we were justified in objecting to the definition

x is true:= x is T

on the grounds of the transition from (1) to (3), then we should be likewise justified in objecting to the definition

x is a bachelor:= x is an unmarried man

on the grounds of the transition from the question

Is Peter a bachelor?

to the following question, which also contains the *concept* of a bachelor (by way of synonymy):

Is Peter an unmarried man?

But we are not so justified; for when we replace a definiendum in a sentence by its (synonymous) definiens, we *expect* no more and no less than that the corresponding *concept* remains the same despite the replacement of the definiendum (the linguistic expression).

8.3.3 Greimann reconstructs Frege's circle objection as follows. He begins with a non-analytic definition:

(W) [The statement] A is true if and only if A has the characteristic [singular!] T.

According to Greimann, the circle is then revealed in the following way:[2]

(A) In applying definition (W) to determine whether a given statement A is true we need to determine whether A has the characteristic T. (assumption)

(B) The question then arises whether it is true that A has characteristic T. (assumption)

(C) To determine whether A has characteristic T we need to determine whether it is true that A has characteristic T. (conclusion
from (A) and (B))

(D) In applying definition (W) to determine whether a given statement A is true we have to already know what truth is. (conclusion
from (A) and (C))

The structure of this argument is somewhat concealed by the fact that
the expressions 'true' and 'T' are used in different syntactical construc-
tions. Thus, 'x is true' is paired with 'x has the characteristic T'. If we
adjust the syntax of the two expressions in this context then we obtain
the following reformulation of this argument:

(W*) It is true that A if and only if it is T that A.

(a) In applying definition (W*) to determine whether it is true that A,

we need to determine whether it is T that A.

(b) The question then arises whether it is true [= T] that it is T that A.

(c) To determine whether it is T that A,

we need to determine whether it is true [= T] that it is T that A.

Conclusion from (a) and (c):

(d) In applying definition (W*) to determine whether it is true
that A,

we have to already know what truth [= T-hood] is.

Greimann's definition (W) or (W*) exemplifies precisely the non-analytic
type of definitions that I discussed earlier. Therefore, my critical objections to the use of such definitions in an attempt to reconstruct Frege's
circle objection likewise apply in Greimann's case. I should like to add
the following additional considerations:

I. Each of the five lines in (a) – (c) contains at least one of the truth
predicates 'true' and 'T', which are synonymous based on the definition. Each of (b) and the second line in (c) even contains both truth
predicates. But nowhere does this argument attempt to 'reduce truth
to something simpler', for *T-hood* is not by any means analyzed into

simpler components than *truth*; it is identical to the latter concept. Yet Frege's result of the circle objection in 'Logik' was this:

(L12) Truth is obviously something so primitive and simple that it is not possible to reduce it *to anything still simpler* [my emphasis].

Note also the plural in

(G9) For in a definition certain characteristics [!] would have to be specified.

For this reason, I believe that Frege's circle objection is directed against analytic definitions such as 'x is true:= x is A and x is B', but not against non-analytic definitions such as 'x is true:= x is T' – not, in other words, against definitions of the kind employed in Greimann's reconstruction.

II. If Greimann's reconstructed circle objection were sound, then *in principle* no expression whatsoever could be synonymous with 'true'. Definitions such as these:

Proposition x is true:= proposition x is correct

It is true that p:= it is a fact that p

have precisely the same structure as Greimann's definition (W) or (W*); therefore, his circle objection would apply to them as well. Consequently, these definitions would be ruled out as faulty and neither of these expressions ('x is correct' and 'x is a fact') could be regarded as synonymous with 'x is true'. But why should we accept this?

Furthermore, the second of the two definitions above appears to be supported by Frege himself, when he writes: 'A fact is a thought that is true' (Frege: 1918a, p. 74).

Finally, we may want to introduce a new term by stipulating:

x is P:= x is true,

and later, once the term 'P' has been established, reverse the definition and consider that

x is true:= x is P.

I do not see how we can deny the legitimacy of such non-analytic definitions. Though such definitions do not contribute in any way to conceptual analysis, their application does not create a circle. When we apply such a definition we merely replace one term by the other – and there is nothing problematic about such a procedure.

III. The crucial step in Greimann's circle objection is the transition from the indirect question in the first line to the indirect question in the second line of (c):

> whether it is T that A,
>
> whether it is true that it is T that A.

If I am correct, this transition does not indicate a circle. However, it could be regarded, with Dummett and Soames, as the beginning of an infinite regress whose next step would look like this:

> whether it is true that it is true that it is T that A.

In short: if we fail to notice that Frege's circle objection is directed specifically against an *analytic* definition, the application of which enables the *decomposition* of a complex initial question into several simpler subquestions, this will substantially affect our reading of the objection.

8.3.4   Kutschera follows Dummett in his interpretation of Frege's circle objection, and in doing so commits precisely the same error as Greimann. He starts out by fully quoting the third paragraph of Frege's essay 'Der Gedanke' and subsequently comments on this paragraph as follows:

> States of affairs [= thoughts] can be called true, according to Frege, only if they correspond to facts. But facts are true states of affairs, and thus the question of being true gets reiterated ad infinitum. A. Tarski specified the truth predicate with regard to sentences such that a sentence is true iff the state of affairs expressed by that sentence exists. Likewise, we could explain the predicate in application to states of affairs [= thoughts] as follows: [(A)] A state of affairs is true iff it [the state of affairs] exists. This explanation is correct, since the expression 'true' does not occur in the definiens. (Kutschera, 1989, pp. 186–7)

As we can see from the quoted passage, according to Kutschera the lack of soundness in Frege's circle objection is immediately based on

Tarski's definition of truth (or in any case, a definition of truth ascribed to Tarski). I shall here refrain from going into whether the first two sentences of the above quote indeed qualify as an abbreviated version of Frege's argumentation (see Chapter 10). But either way, Kutschera's attempt to refute Frege's circle objection by proposing the circle-free definition (A) is unsuccessful.

First of all, Frege does not claim that every *definition* of truth is circular but that every such definition's *application* is. This is why a proof that the definition is 'correct, since the expression "true" does not occur in the definiens' does not in any way affect the soundness of Frege's objection against the definability of truth.

Second, the definition '(The state of affairs) x is true:= (the state of affairs) x exists' is akin to 'Aubergine:= Eggplant'; it is not an analytic definition. Frege's thesis that truth is indefinable, however, means that with respect to truth 'it is not possible to reduce it to anything still simpler' (Frege, 1897, p. 140 (129)), and the above definition is not a counterexample to this thesis.

In a nutshell, Kutschera's critique amounts to the following claim: Frege overlooked that a definition of the type 'Aubergine:= Eggplant' is non-circular. With that Kutschera's critique once again illustrates what is common in (almost) all interpretations dealing with Frege's objections to the correspondence theory and to the definability of truth: they take it for granted that, first, Frege's objections can be quickly dismissed by reference to Tarski's definition of truth, and second, it can be thereby shown just how obtuse Frege's objection is.

8.3.5 Though some scholars clearly notice that Frege has an analytic definition in mind and consequently use an analytic definition such as 'x is true:= x is A and B' in their reconstruction of the circle, they do not employ the step of decomposing a complex question into simpler subquestions, which is made possible by the very analytic nature of such definitions.[3] But if we merely replace the definiendum 'is true' by the definiens 'is A and B' then we could just as well use the shorter 'is T' in place of 'is A and B'. The circle objection, however, will not work *without* decomposition into subquestions.

Perhaps Frege himself is in part to blame for the fact that to date almost all scholars have overlooked this step of decomposing an initial, complex question into simpler subquestions. For in his drafts of the circle objection Frege himself dispenses with an exposition of this decomposition process. Yet, there must be a reason why he makes his claim about the indefinability of truth only with regard to analytic definitions. What explains the difference between applications of analytic and of non-analytic definitions, respectively? Only applications of analytic definitions enable a decomposition of more complex questions into less complex subquestions. This is why the decomposition step is essential in any appropriate reconstruction of Frege's circle objection – and once the step is included in the reconstruction, the circle objection becomes a sound and convincing argument.

8.4   Künne's critique

8.4.0   My reading of Frege's circle argument presented above in section 8.2 was presented to a workshop on Frege at the University of Leiden (Netherlands) in August 2001. Wolfgang Künne presented a similar reconstruction of Frege's circle argument in his 2003 book *Conceptions of Truth* (Künne, 2003, pp. 132–3). In that book, however, Künne goes on to raise the following objection:

> Is this argument convincing? One problem lurks in the presupposition behind (df) [(df) $\forall x$ (x is true iff (x is δ_1, and x is δ_2))]. An analytic definition need not have this structure, and clearly 'x agrees with reality' does not have it. But Frege's vicious circle argument is supposed to cover it as well as any other candidate for the role of an analytic definiens of 'true', whatever its structure. So his talk of 'marks' ['Merkmale', 'characteristics'] in the passage under discussion can amount to no more than the requirement that the alleged definiens is *not atomic* (unlike that in 'For all x, x is a serpent iff x is a snake'). [...] It is far from clear how the argument would have to be modified in order to obtain the required scope. (Künne, 2003, p. 133)

Unfortunately, Künne does not say why he thinks that Frege claims such comprehensive applicability for his circle objection in 'Der Gedanke'. But the way Künne understands Frege's concept of a characteristic or mark is incompatible with comprehensive applicability:

> According to Frege's official use of this word, *M* is a mark of the concept expressed by the predicate 'F' iff the following condition is fulfilled: *M* is a concept, *M* is expressed by a component of an analytic definition of 'F', and nothing can fall under the concept *F* without falling under *M*. This condition is met if and only if the definiens of 'F' is *conjunctive*. Thus, in virtue of the definition 'For all x, x is a

drake iff (x is a duck and x is male)', the concepts expressed by 'male' and by 'duck' are marks of the concept expressed by 'drake'. (Künne, 2003, p. 132)

If Frege understood 'mark' in this sense and at the same time claimed comprehensive applicability for his circle objection, he would be contradicting himself.

8.4.1   In Künne's view, the circle argument, though correct with respect to definitions possessing a *conjunctive* structure such as

x is true:= x is A and x is B,

in which truth is defined by means of *characteristics* (or *marks*) A and B, is not applicable to all analytic definitions; in particular, it is not applicable to a definition such as

x is true:= x agrees with reality.

Künne's objection concerns the *scope* of Frege's circle argument. I shall distinguish three questions one could ask about this scope:

(1) What is the intended scope of Frege's circle argument in 'Logik'?
(2) What is the intended scope of Frege's circle argument in 'Der Gedanke'?
(3) Is there a way in which we could generalize the circle argument so that it can be made to apply to all analytic definitions of S-truth?

I shall answer the first two questions in this section and the third one in section 8.5.

8.4.2   In 'Logik', Frege first claims that the circle can be revealed in applications of the following definition:

(L2) [...] An idea is true if it corresponds to reality.

He subsequently generalizes from this and raises the circle objection for every definition of the following type:

(L9) *A* is true if and only if it has such-and-such properties or stands in such-and-such a relation to such-and-such a thing.

With this definition Frege does not state explicitly what *truth bearers* he has in mind here. We should certainly expect that 'A' is intended to refer at least to ideas, since (L9) is a generalization of (L2); perhaps the scope of reference also includes thoughts, which he identifies as the proper truth bearers a little later in the text of 'Logik'. Now, given the *structure* of the definitional scheme above, (L9) should be read as applying to all analytic definitions. For any such definition contains at least two one-place predicates or at least one two-place predicate, and all of those definitions are covered by (L9).

Thus, in 'Logik' Frege does claim precisely what Künne demands – though does not consider it possible – of the circle argument, namely, that it should apply to all analytic definitions of truth.

Künne's counterexample to Frege's circle argument in 'Der Gedanke' is this:

x is true:= x agrees with reality.

The logical structure of this definition coincides with that of Frege's definition in 'Logik':

(L2) [...] An idea is true if it corresponds to [= agrees with] reality.

Given that Frege explicitly claims that we should be going around in a circle when applying this definition, we should expect that he did indeed consider it possible to generalize the circle argument accordingly – or at least he did so at the time he wrote 'Logik'.

8.4.3   In 'Der Gedanke', Frege dedicates his first three arguments to a critique of the correspondence theory of truth as applied to ideas. Now, if truth is defined as correspondence, then it is not defined by means of characteristics – that is, not by means of two one-place predicates – but rather by means of (at least) one two-place predicate. As we know, Frege objects to this relative (relational) concept of truth in his first argument on the grounds of the absoluteness of scientific truth. And subsequent to the third argument, Frege already anticipates the result of the fourth argument, before presenting the argument itself, as follows:

(G8b) But likewise, any other attempt to define truth also breaks down.

These *other* attempts of which he speaks here differ from the previous one both with regard to the *truth bearers* and with regard to the logical structure of the definiens (see for the following section 7.6). Now, I-truth has been ruled out at that point, after the first three arguments, as a serious candidate for scientific truth; hence, ideas are no longer considered as truth bearers. Therefore, only S-truth remains as scientific truth and only thoughts can eventually be considered as proper truth bearers. Since scientific truth is absolute it cannot be relational; hence, it cannot be defined by means of a two-place relation. Thus, in the comprehensive definitional scheme of 'Logik'

(L9) *A* is true if and only if

[(a)] it has such-and-such properties or

[(b)] stands in such-and-such a relation to such-and-such a thing

part (b) is now out of the question as a definition of S-truth. Only part (a) remains:

(L9a) [A thought] *A* is true if and only if [A] has such-and-such properties.

I think it is for this reason that after (G8b) Frege continues in 'Der Gedanke' as follows:

(G9) For in a definition [*of this other kind*] certain characteristics would have to be specified.

8.4.4 If, however, we wanted to read (G9) just like

(G9*) For in a[ny analytic] definition certain characteristics would have to be specified,

then we could use this reading to justify Künne's thesis that Frege does claim comprehensive applicability for his circle argument. However, this would mean ascribing to Frege an extremely poor short-time memory since, after all, he does discuss analytic definitions that do not contain characteristics in his first three arguments.

(G9) essentially is a paraphrase of (L9a). According to (G9) the circle argument is supposed to apply only to definitions by means of characteristics, and no longer to definitions containing two-place predicates (see (L9b)). Furthermore, in 'Der Gedanke' Frege gives up the circle objection that he has explicitly raised in the 'Logik' version of the third argument against the correspondence theory of truth; at the same time, he modifies the third argument substantially. This provides us with some evidence that Frege in 'Der Gedanke' no longer endorses the claim about the applicability of the circle argument to *any* analytic definition of truth.

But if we referred the word 'so' in

> (G12) *So* it is likely that the content of the word 'true' is sui generis and indefinable,

exclusively to the fourth objection, then this would also support the thesis that Frege saw his circle argument as having comprehensive applicability. However, in this reading Frege would offer *two* formulations of the conclusion of the short fourth argument, namely in (G8) and in (G12), while offering only *one* (cf. (G7)) for his first three arguments (which together are almost five times as long as the fourth argument). This inconsistency would be odd. This is why it would be more plausible to read the word 'so' as referring not merely to the fourth argument but to all four arguments in the third paragraph (see section 3.2).

8.4.5   We can, of course, still ask whether the circle argument is generalizable so as to cover all kinds of analytic definitions of S-truth[4] independently of the question of what scope Frege in fact ascribes to his circle argument in 'Der Gedanke'. Künne, however, considers this as rather unlikely, for he writes: 'It is far from clear how the argument would have to be modified in order to obtain the required scope' (see above). Certainly, in 'Logik' Frege did not consider it to be 'far from clear' how the circle argument reconstructed above could be generalized.

8.5   A generalization of the circle objection

8.5.0   If we wish to generalize from the conjunctive structure of the definition

x is true:= x is A and x is B

to all analytic definitions of truth, we can then characterize the application circle in general as follows:

(1) Due to the analytic nature of the definition, the initial question 'Is it true that p?' can be answered after first answering certain other questions 'q1?', 'q2?', ... (or, more precisely, at least one of these questions).

(2) The interrogative sentences 'q1?', 'q2?',...do *not* contain a truth predicate.

(3) In accordance with (F*), each of the interrogative sentences 'q1?', 'q2?',...is synonymous with one of the following interrogative sentences, which do contain truth predicates: 'Is it true that q1?', 'Is it true that q2?',...

(4) In applying the truth definition we thus need to answer questions which presuppose our knowledge of the concept of truth.

I would now like to show that this type of circle can be revealed in the applications of all analytic definitions of truth. As we can easily see, every analytic definition belongs to one of the following types:

(a) The definiens contains at least two one-place predicates and one sentential operator '#':

x is true:= x is A # x is B.

(b) The definiens contains at least one two-place predicate:

x is true:=...x stands in relation R to α.

Here 'α' may be either a bound variable or a constant. If 'α' is a bound variable then the dots '...' within the definition are replaced by a quantifier, but if 'α' is a constant then the dots are not replaced by anything else; they are eliminated.

In section 8.2 I argued that in any analytic definition of type (a) the circle can be shown to emerge in the applications of that definition. Thus, it remains to be shown that the circle can also be proven to occur in connection with any definition of type (b).

8.5.1 I shall start with a definition that contains a two-place predicate and the *existential quantifier*:

x is true:= There is a y such that: x stands in relation R to y.

In applying this definition we presuppose a range of individuals {a1, a2,...}. We now ask: is a given thought c true according to this definition? This depends on how we answer the following questions:

Does c stand in relation R to a_1?

Does c stand in relation R to a_2?

... ...

None of these interrogative sentences contains a truth predicate, but according to (F*) each of them is synonymous with an interrogative sentence of this type:

Is it true that c stands in relation R to a_1 (a_2, ...)?

To answer this question we must already have the concept of truth. Only after answering this type of question in the affirmative at least once, for example with

It is true that c stands in relation R to a_5,

can we say:

There is a y (namely, a_5) such that: c stands in relation R to y. Hence c is true in accordance with the definition above.

Thus, we must already have the concept of truth in order to be able to determine that c satisfies the condition formulated in the definiens ('There is a y such that: x stands in relation R to y.').

To make this point clearer, let us look at an example. Künne's 'modest account of truth'

– x is true $\leftrightarrow \exists p$ (x = [p] & p) –

is intended as 'the *minimal definition* of (propositional) truth' (Künne, 2003, p. 337). The expression 'x = [p]' here means the same as 'x = the proposition that p'. If we wish to know whether, for example, Hausdorff's maximal principle is true according to this definition, then we select the set of all mathematical propositions $\{p_1, p_2, ...\}$ as our range of individuals and ask which of these propositions is identical to Hausdorff's maximal principle. Once we apply (F*) we have thereby shown the circle.

8.5.2 Analogously we can argue, if we replace the existential quantifier in the above definition by the *universal quantifier*; then we obtain:

x is true:= For all y: x stands in relation R to y.

And here, too, we have to first answer the questions

Does c stand in relation R to a_1?
Does c stand in relation R to a_2?

... ...

which then reveals the circle with the help of (F*) just as in the earlier case above.

8.5.3   If the definition contains a *constant* 'α' in place of the bound variable then it has the following form:

x is true:= x stands in relation R to α.[5]

Here, too, we need to assume a range of individuals {a1, a2, ...}. And in applying this definition, the following questions must first be answered:

Is $\alpha = a_1$?
Is $\alpha = a_2$?

... ...

According to (F*), these interrogative sentences are synonymous with interrogative sentences containing the truth predicate. This proves once again that there is a circle in the application of the definition.

With that it has been shown that the circle emerges in the application of *any* analytic definition of truth (as applied to thoughts).

9
The Omnipresence of Truth

9.1   The circle in the application process

Dorothea Lotter commented on my interpretation of the fourth argument in my earlier book *Freges Kritik an der Korrespondenztheorie der Wahrheit* that since Frege does not explicitly distinguish between two kinds of circularity, reading such a distinction into the intentions of the author may require a fairly generous application of the principle of charity.[1] In this section I shall argue for the presence of a distinction between circular *definitions* and circular *applications* in Frege's text on the basis of that text. I believe that such a justification is possible if we keep in mind the notion of the omnipresence of truth, which I have used above as Frege's presupposition in my reconstruction of the third and fourth arguments.

The following four quotations (with my emphases) are from the two longish passages from 'Logik' and 'Der Gedanke', respectively, that were quoted in section 7.1. In 'Der Gedanke', Frege writes in the context of the fourth argument that 'in *application to any particular case* it would always depend on whether it were true that [...].' This is similar in content to what he writes in the context of the third argument: 'And what would we then have to do so as to *decide* whether something were true? We should have to *inquire* whether it were true that [...]'.

Though Frege uses the word 'application' only in the context of the fourth argument, applications of a definition of truth are also at stake in the third argument; this becomes more obvious if we paraphrase the respective parts of the argument (see (vi)) as follows:

Whenever we want to *decide*
whether (1) something [an idea] were true,

118

and for this purpose *inquire*

> whether it were true that (2) [this] idea and something real correspond in the specified respect [H*],

we apply definition (D_3) to a particular case; for we derive sentence (2) from (1) by substituting the definiens for the definiendum, which is obviously an application of the definition. Thus, with regard to subject matter, both arguments deal with applications of definitions to particular cases.

The parallel passage in 'Logik' likewise focuses on the application of the definition in a particular case. Here Frege writes in the context of the third argument (see section 7.1),

> Now it would be futile to employ a definition in order to make it clearer what is to be understood by 'true' [...] since in order to *apply* this definition we should have to *decide* whether *in a particular case* an idea did correspond to reality [...].

Frege here does not point to a *flaw* (such as circularity) in the definition itself, but rather to the futility *of employing it*, which is revealed whenever we attempt to apply the definition in a particular case.

Finally, Frege also emphasizes the issue of application in a particular case in the 'Logik' version of the fourth argument: '*In each case in hand it would always come back to the question* [...].'

Therefore, I think we can confidently assume that both arguments, in both of their versions ('Logik' *and* 'Der Gedanke'), were intended to reveal a circle specifically in the *application of a definition* to a particular case.

9.2   Propositional questions (1)

The third and fourth arguments clearly differ from the other arguments with regard to a certain stylistic device: the propositional question. To show this I shall quote once more the third argument in 'Der Gedanke':

> Or does it? Could we not maintain that there is truth when there is correspondence in a certain respect? But which respect? And what would we then have to do so as to decide whether something were true? We should have to inquire whether it were true that, for

example, an idea and something real correspond in the specified respect. And with that we should be confronted again by a question of the same kind, and the game could start all over. (vi)

While propositional questions (both direct and indirect ones) are entirely missing in the first two arguments, the third one starts out with no fewer than six interrogative sentences, among which are four propositional questions (at least, if we regard the two indirect questions 'whether...' [= 'Is...this the case?'] as propositional questions). We also encounter a(n indirect) propositional question in the fourth argument (third paragraph):

[...] whether it were true that the characteristics were present. (vii)

The employment of this device creates a clear stylistic break between the third and fourth arguments, on one hand, and the first two arguments, on the other. Another difference consists in the fact that only the third and fourth arguments presuppose the omnipresence of truth. Could there be a connection between these two distinguishing features of the third and fourth arguments? Are propositional questions somehow intrinsically linked to the omnipresence of truth?

Frege begins his essay 'Die Verneinung' with the following paragraph:

A propositional question contains a demand that we should either acknowledge the truth of a thought, or reject it as false. [...] The answer to a [propositional] question [...] is an assertion based upon a judgment; this is so equally whether the answer is affirmative or negative. (Frege, 1918b, pp. 143–4)

'Der Gedanke' contains a similar passage on propositional questions about two pages after Frege's critique of the correspondence theory:

[After asking a propositional question we] expect to hear 'yes' or 'no'. The answer 'yes' means the same as an assertoric sentence, for in saying 'yes' the speaker presents as true the thought that was already completely contained in the interrogative sentence. This is how a propositional question can be formed from any assertoric sentence. [...] An interrogative sentence and an assertoric one contain the same thought; but the assertoric sentence contains something else as well, namely assertion. The interrogative sentence contains something more too, namely a request [that is, the request 'that we should either acknowledge the truth of a thought, or reject it as false' (added from the previous quotation)]. (Frege, 1918a, p. 62)

Thus, to affirm a propositional question means to 'acknowledge as true' the thought expressed by the sentence, while to negate the propositional question means to 'reject it as false [= not true]'. The *request* or *demand* contained in a propositional question can be understood and responded to only if we already have the concept of truth (or falsity (= non-truth)).

What words we should choose in order to express our acknowledgment of the thought as true remains open in these quoted passages, but it seems plausible that Frege had the following types of formulation in mind: either we may simply say 'yes', or 'it is/does', or 'it is true that...'. The accumulation of propositional questions in the third and fourth arguments points to the fact that we all need the concept of truth in order to appropriately engage in the use of propositional questions and their answers.

9.3   Decisions and applications

Frege's first two arguments aim at showing that the respective *definientia* of their targeted definitions of truth do not satisfy the requirements of scientific truth. That is to say, Frege's critique in these arguments is based on the contents of the respective definitions: scientific truth cannot be adequately defined in terms of correspondence because, unlike truth as correspondence, it is absolute and perfect.

Contrary to this, the third and fourth arguments point to the fact that in applying a definition in a particular case we will have to *decide* how to answer the following propositional questions:

Does picture b correspond to the depicted object in respect to H*?

Is it A that it is raining?

Is it B that it is raining?

Now in order to make these decisions we need to have the concept of truth already; otherwise we would not know what is being asked and, consequently, not be able to answer. This is why the concept of truth is omnipresent: By asking 'Is/does it?' I raise the same question as in 'Is it true that it is/does?'

In the above reconstructions of the third and fourth arguments, I have given the respective questions of the form 'Is it?' center stage and subsequently applied Frege's presupposition of the omnipresence of truth, namely (F*), to them. Now, however, this presupposition itself can be justified on the basis of the propositional questions and their context.

In the third argument we have to 'decide' or 'inquire' (vi) whether the idea corresponds to the depicted object with respect to H*. But according to Frege I can understand this very propositional question only if I understand it as the following question: 'Is it true that the idea corresponds to the depicted object in respect H*?' And this shows that I-truth is dependent on S-truth, which is why it cannot be the same as the (independent) scientific truth.

The crucial sentence in the fourth argument is this: 'And in application to any particular case it would always depend on whether it were true that the characteristics were present' (vii). Even if truth is defined by characteristics A and B, the usefulness of this definition depends on its successful *application* in particular cases, that is, on whether I can answer propositional questions of the following kind:

Is it A that it is raining?

Is it B that it is raining?

But in order to raise or answer these propositional questions, we already need to have the concept of truth available, as the more explicit paraphrases of those interrogative sentences reveal:

Is it true that it is A that it is raining?

Is it true that it is B that it is raining?

The omnipresence of truth is revealed in the third and fourth arguments only in the propositional *questions* that need to be answered in the process of applying the definitions to a particular case; it is not revealed in the *definiens* per se. In contrast to his procedure in the first two arguments, Frege here does not criticize the definiens, but instead shows what happens whenever we attempt to apply it.

Lotter is certainly right when she writes that Frege does not explicitly distinguish between two kinds of circularity. For Frege does indeed not explicitly distinguish between a circular *definition* and a circular *application* of a definition. Yet the omnipresence of truth does not enter via the definiens, anyway, but instead via the propositional questions that are required in the process of applying the definitions. This is why I believe it does not require a fairly generous application of the principle of charity for us to attribute to Frege a distinction between circular definitions and circular applications on the basis of Frege's talk of the 'application in a particular case'.

9.4 Propositional questions (2)

The fact that Frege here, in the third and fourth arguments, makes such systematic use of propositional questions as a stylistic device suggests that he may have made such use of them also in the one and only other passage in the third paragraph where he employs them at all. This passage is even earlier than the first argument and is embedded in the following context:

> We find truth predicated of pictures, ideas, sentences, and thoughts. It is striking that visible and audible things occur here along with things which cannot be perceived with the senses. This suggests that alterations in sense have taken place. Indeed they have! For is a picture, as a mere visible and tangible thing, really true? And a stone, a leaf is not true? (i)/(ii)

Here, the *content* of the first question concerns the I-truth of pictures, and yet the *form* of the propositional question introduces S-truth as well: we can raise and answer the propositional question regarding the I-truth of pictures only if we already have the concept of S-truth available. Thus, as early as in this passage, Frege indicates by the form of his propositional question that S-truth is indispensable in science; although at that point he does not explicitly say so, he does already show it via his formulation. For whatever our question is about – let it be I-truth – we must always have the concept of S-truth in order to be able even to ask a propositional question. Thus the solution to the problem of 'alterations of sense' is already implicit in this (and the subsequent) propositional question: S-truth is fundamental, and I-truth must be reduced to S-truth.

It is possible to ask again at this point whether I might not be over-interpreting Frege's text. I do not believe so; but even if this were so, I would not mind. I would consider it a minor sin compared to the 30-year predominance, in the aftermath of Dummett's work on Frege (see Chapter 10), of reading the sentence 'So we should be going round in a *circle'* (vii) unscrupulously as 'So we should be entering an *infinite regress.'*

This interpretation amounts to nothing less than a violation of the text and exemplifies the widespread contempt toward Frege's own actual words. Just as forced and absurd are the erroneous readings of Frege's first argument (see section 4.3). In overcoming this contemptuous mood it will do no harm to, at least tentatively, take Frege's own words seriously to the extent that we may take even the forms of his sentences into account in order to arrive at a fair reading of his text.

9.5   The form of assertive sentences

Only a few pages after his critique of the correspondence theory, Frege develops a systematic connection between thoughts, questions, assertions, and 'that sort of truth which it is the aim of science to discern'. Since these considerations are essential for his thesis of the omnipresence (or redundancy) of truth I shall quote this passage here:

> It is also worth noticing that the sentence 'I smell the scent of violets' has just the same content as the sentence 'it is true that I smell the scent of violets'. So it seems, then, that nothing is added to the thought by my ascribing to it the property of truth. [...]
>
> In order to bring out more precisely what I mean by 'a thought,' I shall distinguish various kinds of sentences. [...] We should not wish to deny sense to a command, but this sense is not such that the question of truth could arise for it. [...] Propositional questions [...] are a different matter. We expect to hear 'yes' or 'no'. The answer 'yes' means the same as an assertoric sentence, for in saying 'yes' the speaker presents as true the thought that was already completely contained in the interrogative sentence. This is how a propositional question can be formed from any assertoric sentence. [...]
>
> [...] We express acknowledgment of truth in the form of an assertoric sentence. We do not need the word 'true' for this. And even when we do use it the properly assertoric force does not lie in it, but in the assertoric sentence-form; and where this form loses its assertoric force the word 'true' cannot put it back again. This happens when we are not speaking seriously. As stage thunder is only sham thunder and a stage fight only a sham fight, so stage assertion is only sham assertion. (Frege, 1918a, pp. 61ff)

Frege here makes a number of positive claims concerning the use of the concept of truth in science. I shall here reiterate an important one of these claims: 'We express acknowledgment of truth in the form of an assertoric sentence. We do not need the word "true" for this.' Dirk Greimann has carefully reconstructed the theory of truth that is suggested here as Frege's 'assertion theory of truth' (Greimann, 2003, pp. 76ff). According to it, scientific truth is acknowledged primarily through assertoric speech acts, and even the secondary attribution of truth by means of the predicate 'x is true' succeeds only if it is performed in an assertion that does not lack the 'requisite seriousness' (Frege, 1918a, p. 63).

It is beyond the scope of this book to investigate whether or to what extent this assertion theory amounts to a consistent and convincing theory of truth; for this I shall refer the reader to Greimann's book. My task here is restricted to showing Frege's critique of the correspondence theory of truth to be internally consistent on the basis of the presuppositions that *he* employed. One important presupposition in the third and fourth arguments is Frege's thesis that truth is omnipresent; and this thesis is based on Frege's account of assertoric sentences and propositional questions. In this book I shall not address this matter in further detail.

10
Dummett's Regress

10.1 Dummett's reconstruction and critique of Frege's arguments

As Dummett's exposition of Frege's argumentation shows, he obviously reads the third and fourth arguments as regress arguments:[1]

> [...] the truth of a sentence or thought cannot [...] be reduced to anything else. If, for instance, the truth of a sentence consisted in its correspondence with something, say W, then, in order to determine whether this correspondence obtained, we should have to enquire into the truth of another sentence, namely 'This sentence corresponds with W'. If the truth of this latter sentence consisted in turn in its correspondence with some further thing, W^*, then, in order to determine its truth, we should have to enquire into the truth of the sentence 'The sentence "This sentence corresponds with W" corresponds with W^*'; and in this way an infinite regress is generated.

Under the influence of Tarski, it may seem like the most reasonable thing in the world to read Frege's objections against the correspondence theory as applying to the correspondence theory of the truth of *sentences*. But doing so means parting with Frege's actual text. Though the two first arguments by themselves can be easily transferred to sentences (see section 14.1), any such all-too-hasty transferral will destroy all access to the third argument, for which the difference between I-truth (in terms of correspondence) and S-truth is essential. If my reading is correct, Frege intends to show here that I-truth, conceived in terms of correspondence, essentially depends on S-truth. This is possible only if the argument deals with *two different* concepts of truth. In Dummett's exposition of the third argument, however, only *one* kind of truth is discussed there: S-truth, construed in terms of correspondence. This is why Dummett cannot interpret the third argument adequately.

Dummett continues:

> The same reasoning shows that truth is absolutely indefinable: for, if the truth of a sentence were to be defined as its possessing such-and-such characteristics, we should have, in order to determine whether the sentence was true, to enquire into the truth of the sentence which ascribed those characteristics to the first sentence; and again we should be launched on an infinite regress.

Again, in the process of application of the analytic definition to a particular case, Dummett's reconstruction of the argument is restricted to the replacement of definiendum by definiens. The reconstruction thus includes no logical decomposition of a complex question into subquestions. This does not do justice to Frege's original argument, in which the decomposition of the truth concept into *several* component characteristics is emphasized.

And here is Dummett's critique of the thus reconstructed argument:

> This argument gives a first impression of sophistry. For, one might say, by this means we could show that the notion of truth had to be rejected altogether, whether defined or not: for the same infinite regress can always be generated. [...] The possibility of the regress thus has nothing to do with whether truth is definable or not. Furthermore, the argument might continue, the regress is not vicious. [...] It is true enough that, in determining that some statement A is true, I thereby also determine the truth of infinitely many other statements, namely 'A is true', 'The statement "A is true" is true', ... But there is no harm in this, as long as we recognize that the truth of every statement in this series is determined simultaneously: the regress would be vicious only if it were supposed that, in order to determine the truth of any member of the series, I had first to determine that of the next term in the series.

Dummett then arrives at the following conclusion about Frege's argument:

> This objection succeeds in showing that Frege's argument does not sustain the strong conclusion that he draws, namely that truth is absolutely indefinable.

10.2 Dummett's and Soames's neglect of Frege's definitions

10.2.0 Dummett's critique of this regress argument, taken by itself, is convincing. It does nothing, however, to undermine Frege's third or fourth argument; all it does is undermine the two arguments as Dummett erroneously interprets them. First of all, neither in the third nor in the fourth argument does Frege speak of a *regress*. True, in the third argument we do encounter the claim '[...] and the game could start all over', but this does not amount to a regress claim. A regress is an *endless* game; it is not a game that begins again. In the context of the fourth argument Frege writes: 'So we should be going round in a *circle.*' We should expect that Frege was aware of the difference between a *circle* and a *regress*. This is also confirmed by what he says in the context of the fifth argument, namely, '[...] otherwise the question of whether something is true would get reiterated to infinity' (x). Here, a regress can indeed be generated, but not one that has anything to do with *Dummett's* regress (see section 12.3).

Dummett is correct when he writes: 'The possibility of the regress thus has nothing to do with whether truth is definable or not' (see above). He is correct, in particular, about the fact that the regress that he, Dummett himself, constructed, has nothing to do with the two kinds of definitions of truth that Frege targets in the third and fourth arguments, respectively. Neither of those kinds of definitions were given any attention in Dummett's construction of the regress; in fact, Dummett could have constructed his regress just as well on the basis of a non-analytic definition of the form 'x is true:= x is T'.

10.2.1 The tendency, already present in Dummett's work, to treat the definitions discussed by Frege as irrelevant is even stronger in Scott Soames's work on Frege. Adopting Dummett's reading, Soames characterizes Frege's third and fourth arguments as follows:

> [...] Frege gives a general argument directed, in the first instance, against all attempts to define truth in terms of correspondence and, by implication, against all definitions of truth whatsoever. (Soames, 1999, p. 24)

Soames then briefly addresses Frege's second argument (see above), quotes the text of the third and fourth arguments, and finally begins his reconstruction of the regress with the following assumption: 'A.

Suppose that truth is definable and that the definition is as follows: For any proposition p, p is true iff p is T' (Soames, 1999, p. 25). With this, Soames merely draws his consequences from Dummett's interpretation. These consequences are correct; for if the structures of the definitions in Frege's third and fourth arguments are irrelevant for the construction of the regress, we need not bother about them and are free to directly construct the regress on the basis of a definition such as

x is true:= x is T.

If the word 'true' can be defined simply as 'T' then we may even use as the basis for our regress a *non-analytic* definition such as

Aubergine:= Eggplant,

which consists in merely replacing one word with another. But this amounts to explicitly disregarding Frege's own text; for Frege speaks in the third argument of truth as a 'correspondence in a certain respect' and in the fourth one of a definition by 'certain characteristics' (in the plural!) – hence in both cases of an *analytic* definition.

10.3 Dummett's way of connecting the third and fourth objections

A crucial part of Dummett's interpretation is the way he reads the connecting sentences between Frege's third and fourth arguments. The last sentence in (vi) and the first sentence in (vii) read,

(a) So this attempt to explain truth as correspondence breaks down.

(b) But likewise any other attempt to define truth also breaks down.

Read from the standpoint of Dummett's exposition, these sentences can only mean,

(a*) So [namely due to *this* regress] this attempt [see (D₃)] to explain truth as correspondence breaks down.

(b*) But likewise [namely due to the *same* regress] any other attempt to define truth also breaks down.

That Dummett's exposition suggests this reading is confirmed by his own paraphrase of (b):

(b_D) The same reasoning shows that truth is absolutely indefinable.

Contrary to this, and as I have shown in detail in sections 3.2 and 7.6, the following alternative reading of (a) does far more justice to Frege's original text:

[(ββ) Result of all *three* initial arguments:] So this attempt [extending from (iv) through (vi)] to explain truth as correspondence breaks down.

Once again, I read the phrase 'this attempt to explain' as encompassing *three different* (though supplemental) definitions of truth as correspondence, where these three definitions are refuted one after the other in *three subsequent* arguments. Based on this reading, the word *'So ...'* in (a) cannot mean 'due to *this* objection (regress)', for at this point Frege really looks back to *three* different objections that he had raised against the various definitions of truth as correspondence. Hence *'But likewise ...'* in (b) cannot mean 'due to the *same* objection (regress) ...'; and therefore Dummett's paraphrase of (b) as

(b_D) The same reasoning shows that truth is absolutely indefinable

should be rejected. Instead, the following paraphrases seem more appropriate:

(a**) So [in the manner as we have just seen] this attempt [encompassing (iv) – (vi)] to explain truth as correspondence breaks down.

(b**) So [that is, likewise, in the same measure (not: based on the same argument)] any other attempt to define truth also breaks down.

10.4   Dummett's regress as a beginner's mistake

10.4.0   In Dummett's regress argument, the concept of truth is applied to *sentences*. According to Dummett we may ignore the particular content of the definition of truth that Frege addresses and instead choose any kind of definition of truth

'x is true:= x is T'.

If we apply this definition to some initial question 'p?', then we obtain with Dummett a sequence of interrogative sentences as follows:

(i) p?

(ii) Is the sentence 'p' true?
(iii) Is the sentence 'p' T?
(iv) Is the sentence 'the sentence "p" is T' true?
(v) Is the sentence 'the sentence "p" is T' T?
(vi) ...

Arguably, not all of the interrogative sentences in this sequence are synonymous; for sentence (ii) presupposes (or implicitly claims) that sentence 'p' exists, while (i) itself does not, and a similar difference can be found in (iii) as compared to (iv); analogously, (ii) and (iv) are not synonymous, and so on (see for this section 12.2). On the other hand, sentences (ii) and (iii), and likewise (iv) and (v) and so on are synonymous because they differ only by the fact that in the second sentence of each pair the word 'true' has been replaced by the synonymous word 'T'. This is why the sequence (i), (iii), (v), ... contains only pairwise different questions – and such sequence of different questions can indeed serve to generate an infinite regress.

10.4.1 In Frege's semantics, however, such a regress would not at all be possible. For, contrary to Tarski and Dummett, Frege applies the concept of truth to *thoughts*, and so his sequence of interrogative sentences would look like this:

(1) p?
(2) Is it true that p?
(3) Is it T that p?
(4) Is it true that it is T that p?
(5) Is it T that it is T that p?
(6) ...

We obtain this sequence of interrogative sentences by alternately applying, on the basis of interrogative sentence 'p?', Frege's presupposition

> (F*) The interrogative sentences 'p?' and 'Is it true that p?' are synonymous.

and the definition of truth

> 'x is true:= x is T'

to each preceding interrogative sentence in this sequence. But both applying (F*) and applying the definition to any preceding interrogative sentence merely lead us – for Frege *obviously* – from one interrogative sentence to another synonymous interrogative sentence, resulting in a sequence of mutually synonymous sentences, all of which express *one and the same question*. Thus, by raising question (1), we have also raised questions (2), (3), and so on, as all of these questions are identical and by answering question (1) we have answered the same question also in its other formulations (2), (3), and so on. This is why the Fregean sequence of interrogative sentences is unsuitable for generating a regress, which would require a sequence of *different* questions.

If Frege indeed thought there was a regress here, he would be caught in a simple contradiction. For on one hand, he would regard interrogative sentences (1), (2), ... as synonymous due to the fact that the sequence had been generated by means of (F*) and the above definition. On the other, he would regard them as non-synonymous, since only in that way could the sequence generate a regress. This would be a beginner's mistake indeed.

10.4.2   We can see at once that the sequence of interrogative sentences consisting of (1), (2), ... is not generated solely by means of the above definition; for if we applied nothing but the definition to (2) 'Is it true that p?' then we would not get beyond (3). That is, the remainder of the sequence could not be generated. (F*), however, would suffice as the sole rule of deduction of an infinite sequence:

(a) p?
(b) Is it true that p?
(c) Is it true that it is true that p?
(d) Is it true that it is true that it is true that p?
(e) ...

As we see here, this infinite sequence is generated by means of (F*) and not by means of the definition. Therefore, this regress, if it were generated, could not serve to undermine the definition. Dummett would have noticed that the two sequences of interrogative sentences (1), (2), ... and (a), (b), ... are equal with respect to the generation of the regress, for he writes: 'The possibility of the regress thus has nothing to do with whether truth is definable or not' (Dummett, 1973, p. 443). But why does Dummett ascribe such an absurd regress argument to Frege? Since this regress has nothing to do with Frege's argumentation I prefer to call it *Dummett's* regress.

10.4.3 When we evaluate an argument, we may not only inquire about its validity or soundness but also about whether it is suitable for reaching its goal. Now would the regress (1), (2),..., if it had been successfully generated, be suitable for Frege's overall argumentative goal?

One implication of the above regress would be the failure of *any* definition whatsoever, not just any definition of truth. For we can reduce any given definition to absurdity in a manner analogous to the one above:

x is a bachelor:= x is an unmarried man.

In applying this definition, we obtain the following interrogative sentences:

(1*) Is Peter a bachelor?
(2*) Is Peter an unmarried man?
(3*) Is it true that Peter is an unmarried man?
(4*) Is it true that it is true that Peter is an unmarried man?
(5*) ...

Thus, in this way any *definition* would break down due to the above regress (1), (2), (3),..., if it could be generated.

Secondly, the regress would cause any given *propositional question* to be unanswerable, since the initial question 'p?' could be any question whatsoever. We are lucky that this argument, whose explosive force is virtually unlimited, is a dud.

Frege develops his third and fourth arguments in order to reduce certain definitions of truth to absurdity and to use these results as contributions to a more solid grounding of logic in the remainder of his essay. Given this ultimate goal in his essay, the infinite regress proposed by Dummett would be just as suitable as an atomic bomb for the purpose of breaking open a safe! If an explosive device does not detonate then the explosives expert has made a mistake. If, however, the explosive device actually has the force to destroy the safe together with all its contents, then the explosives expert would be entirely incompetent. If Frege intended to generate Dummett's regress, he would be making both mistakes.

11
The Reduction of I-Truth to S-Truth

11.1 The text

(viii) When we predicate truth of a picture we do not really mean to predicate a property which would belong to this picture altogether independently of other things. Rather, we always have in mind some totally different object and we want to say that that picture corresponds in some way to this object. 'My idea corresponds to Cologne Cathedral' is a sentence, and now it is a matter of the truth of this sentence. So what is improperly called the truth of pictures and ideas is reduced to the truth of sentences.

(ix) What is it that we call a sentence? A series of sounds, but only if it has a sense, which is not to say that any series of sounds that has a sense is a sentence. And when we call a sentence true we really mean that its sense is true.

[(αα) The answer to the initial question] And hence the only thing for which the question 'Is it true?' can be in principle applicable is the sense of sentences [in short, the thought].

This passage is at the beginning of the fourth paragraph, which previous interpretations of Frege's criticisms of the correspondence theory of truth have almost entirely neglected. But the fourth paragraph is crucial for a correct reading of the third paragraph, since that paragraph makes it very clear that Frege discusses the relation between *two* concepts of truth. In fact, Frege's distinction between I-truth and S-truth is the key to any correct reading of the third paragraph.

11.2 True ideas and true friends

In the third paragraph, Frege has shown that I-truth cannot be scientific truth. It can be concluded from this, within the context of the third paragraph, that S-truth must be scientific truth. But now, in what kind of relation do these two concepts of truth stand?

One possibility is to treat the truth of true ideas as being like the truth of true friends, holding that 'true' as applied to ideas – just like 'true' as applied to friends – is so far from the realm of logic that it becomes irrelevant to Frege's investigation and deserves no further consideration. Frege does, in the second paragraph, set aside the truth of true friends being irrelevant to his concerns.

In the case of the truth of true ideas, however, he proceeds differently, since psychologism is not the kind of opponent that he can afford to simply brush aside and subsequently ignore. As early as in the third argument, Frege establishes a certain relation between I-truth and S-truth: I-truth is dependent on S-truth in its application to a particular case. In the fourth paragraph he spells out the relation between the two concepts of truth more thoroughly, first reducing I-truth to sentence truth, then reducing sentence truth to sense truth (the truth of thoughts). Let us look at these two steps in more detail.

11.3 The reduction of I-truth to sentence truth

Frege reduces I-truth to sentence truth as a first step in passage (viii): I-truth is not the kind of truth that we *'really'* mean when we call something 'true'. This is because I-truth does not belong to ideas 'altogether independently of other things'; 'rather, we always have in mind some totally *different* object'. Truth belongs to an idea only *relative to* this other object; in short, I-truth is relative truth. The idea has to correspond to the other object in order to be I-true, and this is why any correspondence-truth is relative truth. By contrast, *the* truth that we really mean belongs to its bearer 'altogether independently of other things', hence it is *absolute*.

Slightly abridged, Frege's reasoning looks like this:

(A) When we predicate truth of a picture we always have in mind some totally different object,

(B) and we [really] want to say that that picture corresponds in some way to this object.

(C) 'My idea corresponds to Cologne Cathedral' is a sentence, and now it is a matter of the truth of this sentence.

(D) So what is improperly called the truth of pictures and ideas is reduced to the truth of sentences.

The expression 'eigentlich' ('really') actually occurs a little earlier in Frege's text: 'When we predicate truth of a picture we do *not really* mean to predicate a property that [...]. *Rather* [...].' I think, however, that moving the word 'really' in our reconstruction of the argument to a later position does not alter the meaning of what Frege is saying here.

Only after Frege has reduced I-truth to the truth of sentences does he characterize it depreciatingly as 'improper' truth, that is, not the proper truth that he is looking for. But is the truth of sentences already this proper truth? Are sentences the proper bearers of truth?

11.4   The relativity of sentence truth

11.4.0   Frege reduces sentence truth to the truth of senses. This reduction is described rather succinctly thus:

> What is it that we call a sentence? A series of sounds, but only if it has a sense, which is not to say that any series of sounds that has a sense is a sentence. And when we call a sentence true we *really* mean that its sense is true. (ix)

This quote reads like a description of a generally accepted fact; no justification is provided. Obviously, Frege regards it as undisputable that sentences are not properly called true, only their senses, the thoughts they express. That does not mean, however, that he has no reasons for this view. A few pages later we read:

> [It frequently occurs that] the mere wording, which can be made permanent by writing [...], does not suffice for the *expression* of the thought. [...] If a time-indication is conveyed by the present tense one must know when the sentence was uttered in order to grasp the thought correctly. Therefore the time of utterance is part of the *expression* of the thought. If someone wants to say today what he expressed yesterday using the word 'today', he will replace this word with 'yesterday'. Although the thought is the same its verbal *expression* [the sentence] must be different in order that the change of sense which would otherwise be effected by the differing times of utterance

may be cancelled out. The case is similar with words like 'here' and 'there'. In all such cases the mere wording, as it can be preserved in writing, is not the complete *expression* of the thought; the knowledge of certain conditions accompanying the utterance, which are used as means of *expressing* the thought, is needed for us to grasp the thought correctly. Pointing the finger, hand gestures, glances may belong here too. The same utterance containing the word 'I' in the mouths of different persons will express different thoughts of which some may be true, others false. (Frege, 1918a, p. 64; my emphases, modified translation)

To grasp the thought that p correctly, we do not merely need the sentence 'p' but also some 'knowledge of certain conditions accompanying the utterance [in speech or writing]'. Now, if sentences were the proper bearers of truth, then truth would be merely relative, since the same sentence can be true under some conditions but false under others. The sentence, that is, would be true only relative to certain conditions of utterance. Frege already argues in this manner 25 years earlier in his *Grundgesetze der Arithmetik I:*

> And this brings me to speak of what stands in the way of the influence of my book on logicians. It is the corrupting intrusion of psychology into logic. What is crucial to the treatment of the science of logic is the conception of logical laws, and this in turn is connected with how the word 'true' is understood. [...] The sense of the word 'true' could not be more wickedly falsified than by incorporating a relation to those who judge! But surely, it will be objected, the sentence 'I am hungry' can be true for one person and false for another. The sentence certainly, but not the thought, since the word 'I' in the mouth of the other refers to [...] a different person, and hence the sentence uttered by the other expresses [!] a different thought. All specifications of place, time, and so on, belong to the thought whose truth is at issue; *being true* itself is placeless and timeless. [...]
>
> Surveying it all, it seems to me that different conceptions of truth lie at the source of the dispute. For me truth is something objective and independent of those who judge; for psychological logicians it is not. What B. Erdmann calls 'objective certainty' is only a general recognition by those who judge, which is therefore not independent of them but can change [!] with their mental constitution.
>
> We can generalize this still further: I recognize a domain of the objective but non-actual, whereas the psychological logicians

automatically assume that the non-actual is subjective. (Frege, 1893, xiv–xviii)

Stepanians summarizes Frege's position as follows:

> In general, for Frege all parameters that appear to relativize the assignment of a truth-value are but additional specifications of the thought whose truth-value is at issue. In this sense truth is for Frege absolute. (Stepanians, 2001, p. 155)

Since proper truth is absolute, the sentence cannot be the proper bearer of truth, since a sentence's truth is dependent on the conditions of its utterance. Among these conditions Frege lists time, place, and speaker. Another is the *language* to which a sentence belongs; the sentence 'A billion is a thousand millions', for example, is true in contemporary American English but false in (the older) British English.

11.4.1   In contrast to Frege, Tarski proceeds by first selecting the sentence as the proper truth bearer and subsequently arrives at the conclusion that truth is relative:

> The predicate *'true'* is sometimes used to refer to psychological phenomena such as judgments or beliefs, sometimes to certain physical objects, namely, linguistics expressions and specifically sentences, and sometimes to certain ideal entities called 'propositions.' By 'sentence' we understand here what is usually meant in grammar by 'declarative sentence'; as regards the term 'proposition,' its meaning is notoriously a subject of lengthy disputations by various philosophers and logicians, and it seems never to have been made quite clear and unambiguous. For several reasons it appears most convenient to *apply the term 'true' to sentences*, and we shall follow this course. [Footnote (*): For our present purposes it is somewhat more convenient to understand by 'expressions,' 'sentences,' etc., not individual inscriptions, but classes of inscriptions of similar form (thus, not individual physical things, but classes of such things).]
>
> Consequently, we must always relate [!] the notion of truth, like that of a sentence, to a specific language; for it is obvious that the same expression which is a true sentence in one language can be false or meaningless in another.
>
> Of course, the fact that we are interested here primarily in the notion of truth for sentences does not exclude the possibility of a subsequent

extension of this notion to other kinds of objects. (Tarski, 1944, § 2, p. 14)

Tarski here only hints at his reasons for choosing the sentence as proper truth bearer: 'Judgments or beliefs' are *subjective* entities and as such not suitable for logic (a point on which Frege, of course, would agree). Now, for Tarski, propositions – which would roughly correspond to Frege's thoughts – are ideal entities whose ontological status has only led to endless discussions and which are therefore unsuitable to serve as truth bearers. Hence, only sentences remain as suitable truth bearers, for sentences as 'certain physical objects' are neither subjective nor ideal entities and therefore are ontologically unproblematic in Tarski's view.

Whether these reasons are convincing or not is beyond the topic of our discussion here. What is important is the conclusion that Tarski draws from this position: The concept of truth must be 'relate[d] [...] to a specific language' – that means for Frege: it must be relativized.

From a Fregean perspective we can characterize the situation as follows:

(T) If the sentence is the proper bearer of truth then 'truth' is a relative concept.

The logically equivalent contraposition to this implication is:

(T$_C$) *If* 'truth' is not a relative but an absolute concept *then* the sentence is not the proper bearer of truth.

Tarski first selects the sentence as proper bearer of truth and subsequently draws the conclusion from (T) that the concept of truth is to be relativized (in Frege's view): 'Consequently, we must always relate [!] the notion of truth, like that of a sentence, to [!] a specific language.' As the earlier quotes above show, Frege, by contrast, postulates an absolute (and thus non-relative) concept of truth and then uses (T$_C$) to conclude that the sentence cannot be the proper bearer of truth.

11.4.2   I would like to illustrate the above by means of some examples. Concerning a sentence such as

(1) I am hungry. [or (2) A billion is a thousand millions.],

we have generally two possible alternatives. We can say that the *sentence* is true relative to *these* circumstances (such as in this language, for that speaker, ...) and false relative to *those* circumstances; in both sets of circumstances, the sentence (1) [or (2)], conceived of as a certain sequence of letters, is *the same*. Or we can follow Frege and say that the *thought* (A) expressed by the sentence in *these* circumstances is absolutely true, while the *different* thought (B) expressed by it in *those* circumstances is absolutely false.

This is a matter of *strategy*. The sentence

(3) The Republican party's candidate lost the election

can be uttered in many different circumstances C_1, C_2, ... *Either* we use the changed circumstances to relativize its *truth*, in which case we can say that the sentence (3) is (relatively) true in circumstances C_1 but (relatively) false in circumstances C_2. The sentence remains *the same*, though the truth-value of the sentence changes. *Or* we use the changed circumstances to relativize the relation of *expressing*, in which case we can say that the thought expressed by the sentence in circumstances C_1 is (absolutely) true, while the thought expressed by the same sentence in circumstances C_2 is (absolutely) false. In this case, of the two different thoughts one is true and the other one false, and both thoughts retain their respective truth-values no matter how the circumstances may change.

Tarski selects the sentence as the proper truth bearer and consequently relativizes the concept of truth, while Frege starts out by selecting an absolute concept of truth and consequently denies that the proper truth bearer is the sentence. We may choose either of these alternatives, but we should not interpret Frege's argumentation through Tarskian lenses as if he, too, held the sentence to be the proper bearer of truth. Due to his choice of an absolute concept of truth, regarding the sentence as proper truth bearer is out of the question for Frege.

11.5   The truth we really mean

11.5.0   In what follows I should like to take a closer look at the reduction of I-truth to sense truth, since it is the key to understanding Frege's regress argument. This regress argument immediately follows the reduction and concludes the fourth paragraph.

As mentioned earlier, the reduction of I-truth to sense truth consists of two steps, in both of which Frege uses a sentence like

When we call A true we *really* mean that B is true.

He uses this expression in a positive formulation in the last quote concerning his reduction of sentence truth to sense truth: 'And when we call a sentence true we *really* mean that its sense is true' (ix). But the same expression also occurs in a negative formulation at the beginning of (viii):

> When we predicate truth of a picture we do *not really* mean to predicate a property which would belong to this picture altogether independently of *other* things [while by truth, we *really* mean this absolute truth that is independent of other things]. Rather, we always have in mind some totally *different* object and we want to say that that picture corresponds in some way to this object. (viii)

In the context of the third and fourth paragraphs, this quotation indicates that what we *really* mean by truth is absolute truth and that I-truth does not qualify as absolute truth. Subsequently, 'what is improperly ('missbräuchlich') called the truth of pictures and ideas is reduced to the truth of sentences' by Frege. The proper truth of sentences (or senses) is contrasted with the improper truth of ideas, which compared to proper truth is merely an improper kind of truth; this improper truth is then reduced to proper truth. The sense of 'improper' ('missbräuchlich') implies that an improper truth is so to speak an *unreal* truth, but the German word 'missbräuchlich' implicates an even stronger disqualification and rejection of I-truth than the word 'unreal' would. Leaving aside the different colorings of the two expressions, we can summarize the two steps in Frege's reduction as follows:

(R₁) When we predicate (relative) truth of an idea Iy we really mean to predicate (relative) truth of a sentence of the form 'Iy corresponds to Xy [the second relatum]'.

(R₂) When we predicate (relative) truth of a sentence of the form 'Iy corresponds to Xy' we really mean to predicate (absolute) truth of its sense.

11.5.1   If we apply also the second step in Frege's reduction to the example offered in Frege's text then the reduction looks like this:

(I) 'My idea (of Cologne Cathedral) is true'

is reduced to (cf. R₁):

(S) 'The sentence "My idea (of Cologne Cathedral) corresponds to Cologne Cathedral" is true',

and this is reduced to (cf. R_2):

(T) 'The thought that my idea (of Cologne Cathedral) corresponds to Cologne Cathedral is true'.

What we *really* mean when we assert (I) is made explicit in (T). If in (I) we predicate the relative I-truth of an idea, we really mean to predicate the absolute truth of a thought (the sense of a sentence). For what we *really* mean by truth is always the truth of a thought.

The question 'What do we really mean when we call an idea *true*?' focuses on the word 'true'. We expect the answer to contain an explanation or definition of I-truth, which Frege indeed delivers in a concise form: 'we want to say that [that idea] corresponds in some way to this object.' But this is only a tentative answer that misses the proper sense of 'true' even though it is correct in the case of ideas, as we saw from the third paragraph. If we wish to grasp this proper sense then the answer can only be thus: 'When we call an idea *true* we really mean that a *thought* is true.' In this answer, the focus of the question is shifted. The question aims at the sense of the word 'true', while the answer focuses on the *bearer* of truth. This shift of focus is justified in that we are able to grasp the real *sense* of 'true' only by relating it to the real *bearer* of truth. For Frege, the two following questions are closely connected:

What do we really mean by *truth*?

Of what do we really predicate truth?

The first question cannot be positively answered by means of a definition in which we would list the *characteristics* (the content) of the concept of truth. Instead, Frege has to indicate *properties* of truth itself, such as absoluteness, perfection, and independence, in order to specify what truth really is and to distinguish it from I-truth. Since the first question about the content of truth cannot be directly answered, the second question about the proper bearer of truth must take center stage – since this is the question that can be answered.

11.5.2   I-truth, which is *the* truth to be reduced, is relative. Sense-truth, which is the truth *to which* the former is reduced, is absolute. I consider this to be a central point of Frege's argumentation: since we really mean by truth absolute truth, we ultimately relate an assertion in which we predicate relative truth of an idea to *absolute* truth and its *bearer* – both belong together. We cannot predicate absolute truth of an idea because ideas are not true in an absolute sense, they can merely correspond to something. Nor can we predicate the relative truth of correspondence to the sense of a sentence, the bearer of absolute truth, since this would lead us into an infinite regress, as Frege shows with his fifth argument (see Chapter 12). Thus, the different concepts of truth conform to certain suitable truth bearers and vice versa:

(1) Idea – relative truth (correspondence truth),
(2) Sentence – relative truth (depending on the sense expressed),
(3) Sense – absolute truth.

A reduction of one truth concept to another can succeed only because the *truth* to which the other truth concepts are reduced is absolute and because the ultimate *truth bearer* is the bearer of absolute truth. A reduction of truth concepts (together with their corresponding truth bearers) all of which are relative would only lead to an infinite regress.

Frege's basic question in his reduction of truth concepts is:

(R) What do we *really* mean when we call something true?

The answer is: by truth we really mean *absolute* truth, and we really predicate this absolute truth of the senses of sentences, which are therefore the proper bearers of absolute truth.

Concretely, Frege's reduction of one truth concept to another consists in the following theses:

(1) By truth we really mean absolute truth.
(2) We predicate absolute truth of a thought, the sense of a sentence.
(3) The truth of ideas consists in their correspondence to something else and is therefore relative.
(4) When we say that an idea is (relatively) true we really mean that the sentence 'The idea corresponds to something else' is (relatively) true.
(5) When we say that the sentence 'The idea corresponds to something else' is (relatively) true we really mean that the thought that the idea corresponds to something else is (absolutely) true.

11.6   The answer to the initial question

Now Frege is finally able to answer the initial question raised in the third paragraph, namely,

> Of what is scientific truth really predicated? (literally: (α) '[What is] the region within which the question "Is it true?' could be in principle applicable?' (i))

toward the end of the fourth paragraph almost in the same wording:

> [(αα)] And hence the only thing for which the question 'Is it true?' can be in principle applicable is the sense of sentences. (ix)

This question and its answer fuse the third and fourth paragraphs into one argumentative unit. With this answer, the argumentative goal of the two paragraphs has been achieved. Scientific truth, which has been shown in the fourth argument to be indefinable, belongs to the *sense* of a sentence or the thought expressed, and not to an *idea* or picture, whose truth consists in its correspondence to something else.

12
The Fifth Argument: Frege's Regress

12.1 The text

(x) Now is the sense of the sentence an idea? In any case, truth does not consist in correspondence of the sense with something else, for otherwise the question of whether something is true would get reiterated to infinity.

Here, at the end of the fourth paragraph, Frege does introduce a regress argument that – just like his third argument – has never to date even been recognized by scholars as an independent argument. Where it has been recognized at all, it has been perceived as a mere reiteration of an alleged regress argument in the third paragraph – namely, Dummett's regress argument. As I have shown in Chapters 6–8 and 10, however, there is no regress argument in the third paragraph. Therefore, we have to conclude that the present passage contains a completely new argument. And the text does not leave any room for doubt that Frege indeed points out an infinite regress in this fifth argument: '[...] for otherwise the question of whether something is true would get reiterated to infinity.'

12.2 The regress argument

12.2.0 Frege's regress argument has the form of an indirect proof that looks roughly like this:

(A) Assumption: the truth of the sense (of a sentence) consists in its correspondence with something else.

(B) Then the question of whether something is true gets reiterated to infinity.

(C) Therefore, truth of a sense does not consist in its correspondence with something else.

To be shown here is that the regress asserted in (B) follows from (A). Examples for questions about whether something is true are the following:

Is the *sentence* (0) 'Cologne Cathedral has two towers' true?

Is the *sense* S_0 of the sentence (0) 'Cologne Cathedral has two towers' true?

Such questions are supposed to lead to a regress if we assume (A). But (A) alone does not suffice for this, which is why we have to ask for the context in which the regress is to be constructed. Since Frege outlines the regress immediately after his *reduction* of the truth concepts, it is reasonable to reconstruct the regress in that context. That is what I am going to attempt now.

12.2.1   As we saw in Chapter 11, Frege reduces the various concepts of truth to one logically basic one in two steps:

(R_1) When we predicate (relative) truth of an idea Iy we really mean to predicate (relative) truth of a sentence of the form 'Iy corresponds to Xy [the second relatum]'.

(R_2) When we predicate (relative) truth of a sentence of the form 'Iy corresponds to Xy' we really mean to predicate (absolute) truth of its sense.

Frege's reduction ends with the sense as the proper bearer of absolute truth. Since the truth of senses is absolute, it cannot be reduced any further to another kind of truth. Thus the reduction of truth concepts comes to a natural end in the absolute truth of senses.

If, however, the truth of senses according to (A) consisted in a *correspondence to something else* then this truth of senses would be just as relative as the truth of ideas. Assumption (A) relativizes the truth of senses, for according to (A) the sense of a sentence is true only in relation to something else with which it corresponds. When reducing I-truth to sentence truth in (viii) Frege makes use of only a single property of

I-truth, namely the property of its consisting in a correspondence. It is precisely this property that according to (A) would also belong to the truth of senses, and therefore Frege's argument in (viii) can be directly transferred to the truth of senses:

(viii*) When we predicate truth of a [sense] we do not really mean to predicate a property which would belong to this [sense] altogether independently of other things [according to (A), even the truth of senses is not absolute ('altogether independent of other things')]. Rather, we always have in mind some totally different object and we want to say that that [sense] corresponds in some way to this object [or this state of affairs]. '[The sense] corresponds to [X]' is a sentence, and now it is a matter of the truth of this sentence. So what is improperly called the truth of [senses] is reduced to the truth of sentences.

If we now treat the relative truth of senses in the same manner as we previously treated the relative truth of ideas while *ceteris paribus* retaining Frege's reduction of truth concepts we obtain the following two reductive steps:

(R_1*) When we predicate (relative) truth of a sense Sy expressed by a sentence y, we really mean to predicate (relative) truth of a sentence of the form 'Sy corresponds to Xy [the second relatum]'.

(R_2*) When we predicate (relative) truth of a sentence of the form 'Sy corresponds to Xy' we really mean to predicate (relative) truth of its sense.

These steps can reduce a relative truth only to another relative truth. Since, however, according to Frege we cannot stop at a relative truth but need to keep reducing it to another truth, the reduction of truth concepts will lead to an infinite regress, as I shall illustrate in the following.

12.2.2   Here is my reconstruction of the regress. We start out with the question:

Is the *sentence* (0) 'Cologne Cathedral has two towers' true?

Before answering this empirical question with 'Yes' (or 'No') we should clarify in the context of logic what we *really* mean when we call a *sentence* (or its *sense*) true. This logical question of what we really mean pushes us into the regress.

According to (R_2^*) we *really* mean by the relative truth of sentence (0) the following:

   Is the *sense* S_0 of the *sentence* (0) 'Cologne Cathedral has two towers' true?

According to (R_1^*) we *really* mean by the relative truth of the sense S_0 the following:

   Is the *sentence* (1) 'S_0 corresponds to X_0' true?

According to (R_2^*) we *really* mean by the relative truth of sentence (1) the following:

   Is the *sense* S_1 of the sentence (1) 'S_0 corresponds to X_0' true?

According to (R_1^*) we *really* mean by the relative truth of the sense S_1 the following:

   Is the *sentence* (2) 'S_1 corresponds to X_1' true?... ... ...

The truth bearer here changes with each reductive step. Thus, 'the question of whether something is true [gets] reiterated to infinity'. We will never find an ultimate truth bearer Ω of which we could say: when we ask whether the sentence (0) is true, we are really asking whether Ω is true.

In this way, the question of what we *really* mean when we call something true (see section 11.5) leads us into a regress: from sentence (0) we are referred to sense S_0 as the proper bearer of truth, and then from sense S_0 to sentence (1), from sentence (1) to sense S_1, from sense S_1 to sentence (2) and so on ad infinitum. Schematically the process looks like this:

$$(0) \Rightarrow S_0 \Rightarrow (1) \Rightarrow S_1 \Rightarrow (2) \Rightarrow ...$$

Since this approach leads us into a regress, it cannot yield a final answer to the question of what we *really* mean when we call something 'true'. In particular, it cannot yield an answer to the question of what the proper bearer of truth is.

Now, Frege did previously determine the sense of a sentence to be the proper bearer of truth. Therefore, the truth of this sense cannot consist in its correspondence to something else, or else we would enter the regress outlined above.

12.2.3 The result of this reconstruction is an infinite regress only if the sentences (0), (1), (2), ... constitute an infinite number of distinct sentences and their senses S_0, S_1, S_2, ... likewise constitute an infinite number of distinct senses.

If, however, each of the sentences and senses, respectively, in this sequence is identical to the other sentences/senses then we would not obtain an infinite regress but rather a circle. For if the sentences (0), (1), (2), ... were identical to one another and the senses S_0, S_1, S_2, ... were also identical then the above schematization of the regress shows that every other step will get us to where we were before:

$$(0) \; [= (1) =...] \Rightarrow S_0 \; [= S_1 =...] \Rightarrow (1) \; [= (0) = (2) =...] \Rightarrow S_1 \; [= S_0 = S_2 =...] \Rightarrow ...$$

In this case, we do not obtain a regress but a circle or, more precisely, an infinite back-and-forth between the sentence (0) [= (1) = ...] and the sense S_0 [= S_1 = ...].

It remains to be shown that the sentences and their senses, respectively, are indeed distinct from one another. Made explicit, the *sentences* (0), (1), (2), (3), ... look like this:

(0) Cologne Cathedral has two towers.

(1) The sense of '[0]Cologne Cathedral has two towers.[0]' corresponds to X_0.[1]

(2) The sense of '[1]The sense of '[0]Cologne Cathedral has two towers.[0]' corresponds to X_0.[1]' corresponds to X_1.

(3) The sense of '[2]The sense of '[1]The sense of '[0]Cologne Cathedral has two towers.[0]' corresponds to X_0.[1]' corresponds to X_1.[2]' corresponds to X_2.

... ...

The explicit formulation of sentences (0), (1), (2), (3) ... shows that these sentences are pairwise distinct (that is, that each sentence in this sequence is distinct from every other sentence within the sequence).

Hence the sequence (0), (1), (2), ... contains infinitely many distinct sentences.

The distinctness of the *senses* can be justified as follows. The quotation names 'n...n' (with $n = 0, 1, 2, ...$) contained in sentences (1), (2), (3), ... are pairwise distinct and refer to pairwise distinct sentences: the quotation name in (1) refers to sentence (0), the quotation name in (2) to sentence (1), and so on. Since the references of the quotation names are distinct, so too are their senses according to Frege. Now according to Frege's principle of compositionality the sense of a sentence is composed of the senses of its parts. Since the quotation names within the sentences have different senses, the principle of compositionality therefore suggests that sentences (1), (2), (3), ... also have pairwise distinct senses. Furthermore sentence (0) does not contain any quotation name and does not refer to any sentence; in this alone its sense already differs from those of all other sentences in the sequence. Thus, due to the difference in the senses of one or the other of their parts, all senses of sentences (0), (1), (2), (3), ... are pairwise distinct.

Since, however, both the sentences and their senses are pairwise distinct, the schema

$$(0) \Rightarrow S_0 \Rightarrow (1) \Rightarrow S_1 \Rightarrow (2) \Rightarrow ...$$

does indeed represent an infinite regress.

12.2.4   Some additional points should be briefly addressed here.

(a) Would Frege himself say of the *sense* of a sentence that it corresponds to something else? Certainly not, if we consider how strictly he distinguishes in the third paragraph between I-truth and S-truth. But this is not an objection against the regress, for the regress is an indirect proof. Frege's opponent, the psychologist, holds the view that the sense of a sentence is an idea. Frege addresses this thesis as a question: 'Now is the sense of the sentence an idea?' If the sense were an idea then its truth would consist in a correspondence. Frege then assumes for the sake of argument that sense truth consists in correspondence and demonstrates the absurdity of this assumption by showing how it leads to an infinite regress. It follows that the sense cannot be an idea (see section 12.4).

(b) How can we specify the 2nd relata $X_0, X_1, X_{2, ...}$ (see sentences (1), (2), ...) in more detail? I find the answer to this question difficult already with regard to X_1, and I cannot conceive of an answer with regard to X_2. But the question has to be answered, first of all, by those who want to assign a correspondence type of truth to the sense of a sentence – and Frege is not one of them.

(c) To claim that all of the sentences (0), (1), ... are logically equivalent is not a sufficient objection to the regress argument. For the logical equivalence of these sentences does not imply the identity of their senses.

(d) The following consideration suggests, independently and beyond Frege's own argumentation, that the sentences are not logically equivalent. The use of the quotation names $'n\ldots n'$ (with $n = 0, 1, \ldots$) in the sentences $(n+1)$ indicates a presupposition of the existence of the corresponding sentences. Sentences as *sentence tokens* are physical objects just like Cologne Cathedral, and the existence of physical objects is contingent. Therefore, the presupposition that a certain sentence exists cannot be regarded as trivial and always satisfied. If, with Tarski, we regard sentences as *classes* of sentence tokens then we will be able to distinguish these classes only if they are not empty, which is why we presuppose – in using the expression 'the sentence class φ' – that there is at least one token φ. Otherwise, we would obtain absurd consequences. To be sure, the above construction of the sequence of sentences (0), (1), ... ensures that the sentence (n) is produced first, and the sentence $(n+1)$, which contains the quotation name $'n\ldots n'$ for sentence (n) (with $n = 0, 1, \ldots$), is produced only subsequently. Hence none of the quotation names will be empty, that is, these existential presuppositions will always be satisfied. Nonetheless, this does not alter the fact that the sentence tokens are contingent, which contradicts the logical equivalence of the sentences.

12.2.5 We could consider the infinite sequence of sentences and senses to be harmless if only the *verification* of sentences and their senses were at stake. In that case we would not be asking what we really mean when we call something true, but could content ourselves with a vague understanding of truth according to which we can call sentences and their senses true without the need to distinguish them. In effect, we could dispense with the clarification of the concept of truth and attempt to answer the question 'Is this true?' right away with 'Yes' or 'No'. In this case the regress is not at all worrisome. Rather, the situation could be just as Dummett described it in reference to his alleged regress:

> [...] the regress is not [!] vicious. For suppose it truly said that the truth of a statement A consists in its correspondence with some state of affairs W. [...] It is true enough that, in determining that some

statement A is true, I thereby also determine the truth of infinitely many other statements, namely 'A is true', 'The statement "A is true" is true',...But there is no harm in this, as long as we recognize that the truth of every statement in this series is determined simultaneously: the regress would be vicious only if it were supposed that, in order to determine the truth of any member of the series, I had first to determine that of the next term in the series [see section 10.1].

Just as Dummett points out here, the sentences and senses in the sequence

$$(0) \Rightarrow S_0 \Rightarrow (1) \Rightarrow S_1 \Rightarrow (2) \Rightarrow ...$$

could all be true at the same time: By judging that the initial sentence 'Cologne Cathedral has two towers' is true, we have at the same time judged all other sentences and senses in the sequence to be true (though in this case we would be neglecting the fact that according to Frege the word 'true' would be used with different meanings).

Considered in this way, the infinite sequence is harmless; it does not prevent us from *verifying* any randomly picked sentence and its sense. However, before we answer the question 'Is it true?' in a particular case, we should at least in logic answer the following even more basic questions:

What do we really *mean* when we call something true, and *of what* do we really predicate truth?

For such questions it does not help if all sentences and their senses within the sequence are true (whatever truth they may represent). The question about the proper bearer of truth, which is the central question of the third and fourth paragraphs, is indeed blocked by the infinite sequence – here, the regress becomes 'vicious'.

Thus, the infinite regress provides the result intended by Frege: sense truth cannot consist in correspondence with something else.

12.3   On the distinction between circle and regress arguments

12.3.0   In my reconstruction of Frege's arguments 3 – 5 I carefully distinguished between circle and regress arguments. The third argument is neither a circle nor a regress argument, the fourth can be considered a valid circle argument – while Dummett's regress argument, which is supposed to be a reconstruction of Frege's fourth argument, is invalid – and the fifth argument can be reconstructed as a valid regress argument. However, the following objection has been raised against my distinction between circle and regress arguments:

> [Pardey's] analysis and response to critics [of Frege] hinges on a distinction between circular arguments and regress arguments. I do indeed see that Frege uses slightly different language in different places, which tracks this distinction. But most logicians and mathematicians would see no important difference between these two notions. They are equivalent, and play equivalent roles in arguments. The change between the two is so automatic that we might well describe them as notational variants of each-other. To make the close reading interesting, the author would need to substantiate the claim that Frege himself sees a significant difference between these two. I do not see how this case has been made, and the fact that Frege uses slightly different forms of expression does not seem to me to make it. Perhaps more important in response to critics, the author needs to establish that there is some significant logical difference between these two notions, in spite of the general view that there is not. Unless this difference can be established, the response to critics hinges on a distinction without difference, and fails. I do not see how this task can be accomplished, as I do not think there is any difference to be found. But nonetheless, the author needs to explicitly argue there is one in order to make his response to critics compelling.[2]

12.3.1 For several reasons I do not find this objection to my Frege interpretation convincing. First, the *general* claim that circle and regress are mere 'notational variants of each-other' is false, at least with regard to its generality. The difference between these two forms of argument can be summarized as follows: a circle argument aims to show that the *very* concept A that is being defined (or analyzed) is also either explicitly or implicitly contained in the definiens (or analysans), or that the *very* sentence 'p' that is being justified is included either explicitly or implicitly in the justification. Thus the claim is that something is being defined, analyzed or justified *by itself.*

For a regress, by contrast, it is essential that it leads from one sentence to the next, and from there to a third and so on, and that all these sentences are *distinct* from one another. If at any location within the regress there occurs a sentence that has already occurred earlier then we no longer have a regress but a circle. Hence the circle and regress objections are different, easily distinguishable forms of argument.

Now in response to the above quoted claim that there is an argumentatively equivalent regress for every circle, and vice versa, let me draw the reader's attention to a number of regress arguments for which there are no equivalent circle arguments. For example, Aristotle's infinite regress of contingent causes, from which he concludes the existence of a necessary first cause, does not – as far as I can see – have an equivalent circle that could replace it. Nor does Plato's and Kant's postulation of the absolute Good – on the basis of the fact that without the assumption of such a Good we would enter into an infinite regress of merely relative goods (of the type 'x is good for y') – seem to lend itself to reconstruction in terms of a circle instead of a regress. Thus, at least in philosophy, it seems highly problematic to consider circle and regress arguments as mere 'notational variants of each-other' (see above). This is precisely why philosophers commonly distinguish between circle and regress arguments, and this distinction is also shared by Frege scholars such as Künne, Stepanians and Stuhlmann-Laeisz (Cf. Stuhlmann-Laeisz, 1995, p. 26; Künne, 2003, p. 131f, Stepanians, 2001, pp. 157ff).

12.3.2   Secondly, the objection is directed *especially* against my Frege interpretation. Now my interpretation does not rest essentially on the difference between circle and regress arguments *as such*. Instead, it mainly relies on the manner in which Frege's arguments can be reconstructed as *valid arguments*. Dummett's reconstruction of the fourth argument as a regress argument yields an invalid argument. By contrast, my reconstruction of the same argument as a circle argument shows it to be a valid. Now if, as seems to be the above critic's view, circle and regress are mere 'notational variants of each-other', then both arguments should be either valid or invalid, since they both represent reconstructions of one and the same Frege argument. Since, however, Dummett's regress is invalid and my circle argument is valid, the two arguments cannot be equivalent variants of each other. Or else, why did the critic not try to show that, contrary to appearance, Dummett's regress and my circle argument are but equivalent variants of one and the same argument, after all?

Frege's fifth argument has been generally received as an abbreviated version of Dummett's regress, and thus as an invalid argument. My reconstruction of the fifth argument discloses a different regress, which we obtain by asking for the proper truth bearer (see above in more detail) based on the presuppositions (R_1*) and (R_2*):

Is the *sentence* (0) 'Cologne Cathedral has two towers' true?

Is the *sense* S_0 of the sentence (0) 'Cologne Cathedral has two towers' true?

Is the *sentence* (1) 'S_0 corresponds to X_0' true?

... ...

Each sentence in this sequence is distinct from each of the others. Thus I do not see how we could possibly replace this regress by an equivalent circle. Here, too, it would have been helpful if the critic had shown what the alleged circle variant of the regress would look like.

In short: this objection to the common philosophical distinction between regress and circle is poorly justified and unconvincing, especially as an objection ultimately aimed against my interpretation of Frege.

12.4 The context: Frege's major thesis in 'Der Gedanke'

12.4.0 The question

Now is the sense of the sentence an idea?

which Frege, given his rejection of psychologism, certainly intends to answer in the negative by means of his regress argument, is not just any question. Rather, it is the *central* question in the second half of his essay 'Der Gedanke'. It is raised for the first time here on p. 60 of the original German text without being answered directly and is taken up again on p. 66. On pp. 60–66 Frege clarifies in more detail the relation between sentences and thoughts (as senses of sentences); he discusses the traditional distinction between the external world and the inner world and finds it obvious that thoughts do not belong to the *external* world and therefore cannot be identified with sentences. Then Frege asks again:

Now do thoughts belong to this *inner* world? Are they *ideas*? [p. 66, my emphasis]

Again, Frege does not subsequently answer this question directly but instead discusses the question 'How are ideas distinct from the things of the external world?' Then he raises the question a third time:

I now return to the question: Is a thought an idea? [p. 68]

Frege's real answer to this question is a clear 'No!' but the initial formulation of his answer comes across as rather tentative:

> So the result *seems* to be: thoughts are neither things in the external world nor ideas. [p. 69, my emphasis]

This answer is repeated six pages later, *sans* the qualifying 'seems'; here Frege devotes an entire paragraph to it (as he does with the last quotation), because it is the *major thesis* of his essay:

> A thought belongs neither to my inner world as an idea, nor yet to the external world, the world of things perceivable by the senses. [p. 75]

Still, in the next paragraph we find Frege's formulations becoming more tentative again, which goes to show that he anticipates considerable opposition to his positing of a 'third realm' (p. 69), distinct from both the external and the inner worlds, as the proper ontological region of thoughts:

> This consequence, however cogently it may follow from the exposition, will nevertheless perhaps not be accepted without opposition. It will, I think, seem impossible to some people to obtain information about something not belonging to the inner world except by sense perception. [p. 75]

Frege continues this tentative, unassertive tone in later passages:

> Thus I cannot find this distinction [between the ways a thing and a thought are given] to be so great as to make impossible the presentation of a thought that does not belong to the inner world. [p. 75]
>
> A thought, admittedly, is not the sort of thing to which it is usual to apply the term 'actual'. [p. 76]
>
> Thoughts are not wholly unactual [= kind of actual], but their actuality is quite different from the actuality of things. [p. 77]

The regress argument serves to develop an intitial, still hesitant and suggestive, answer to Frege's central question, 'Now is the sense of the sentence an idea?' The regress is meant to show that, in any case, 'truth does not consist in correspondence of the sense with something else'. Since, however, the truth of an *idea* consists precisely in such correspondence, senses systematically lack an important characteristic of ideas, and therefore – as we can conclude already at this point – senses cannot be ideas.

12.4.1   Overall, this exposition has shown *that all five of Frege's arguments are consistent*. We do not find any of the previously mentioned logical beginner's mistakes (see section 1.2) in any of these arguments.

This completes the first part of my interpretation, in which I aimed to show that Frege's arguments are internally consistent. The remaining task is to inquire whether the same arguments are also externally consistent, that is, whether they can be reconciled with widely-accepted core positions of our contemporary views of truth. Before turning to this question I should like to describe some of the stations along the road by which I arrived at my new interpretation.

13
The Road to a Novel Interpretation

Some parts of my interpretation, especially those concerning Frege's first and third arguments, differ so greatly from previous interpretations that the latter do not even offer anything that could provide a departure point for the novel interpretation. Disregarding whether my interpretation is accurate, there is still the question of how one can arrive at such a novel interpretation in the first place. Is it based on a sheer intuition that a person either has or does not have?

Though I will not deny that intuition plays an important part, that part is less important than it may seem to be at first sight. In particular, intuition does not fall from the sky but materializes, if at all, only after intensive work on the text. Arriving at a novel interpretation is, in my experience, a time-consuming and often unclear process. Long into that process we do not know whether the text deserves the effort in the first place, or whether our attempts at an interpretation might not be a dead-end route, after all. For these reasons it is, at the very least, quite risky to engage in developing at a novel interpretation.

This present chapter is dedicated to some of the mileposts on my road to a novel interpretation of Frege's first and third arguments. We will see that the previous interpretations did, after all, contribute to the development of the new one. For students of philosophy eager to learn the art of interpretation, the road by which we arrive at the interpretation will itself be just as important as the results. More advanced Frege scholars, meanwhile, may find this Chapter interesting because it offers some variants of the interpretation presented in previous chapters.

13.1 Contrapositing an interpretation

13.1.0 Nowadays, when we set out to interpret a philosophical text we usually have not only the original text but also other previous interpretations that we can study. These previous interpretations may be obstructive, or helpful, or both.

They can have an obstructive effect if we read them prior to or simultaneously with the original text, because they may affect our understanding and lead to a selective perception of the original text. If we then read the text in the manner suggested by the interpreter, we may not even notice that certain crucial aspects of the text have been entirely neglected or misunderstood by the interpretation, and we end up reading the original text superficially and merely as a confirmation of the interpretation. It is then often difficult to rid ourselves of the influence of this too prematurely read interpretation.

Previous interpretations can be helpful provided we first study the original text; only after that, once we have either acquired our first understanding of the text or found ourselves blocked by major difficulties in understanding it, is it time to turn to them. If we come upon such an interpretation that deviates greatly from our own, we should read the original text once again very thoroughly to find out which interpretation is closer to the text, or whether they are perhaps both equally legitimate. In the latter case, we may also arrive at a third interpretation.

In earlier Chapters I mainly attacked the previous interpretations, rejecting them as inadequate. In this Chapter, by contrast, I am going to show just how these other interpretations assisted me in arriving at my own.

13.1.1 My novel interpretation of Frege's theory of truth owes at least as much to my critical study of previous interpretations by Dummett, Künne, Soames, Stuhlmann-Laeisz and so on as it does to my own reading of Frege's original text. I frequently applied the rule of contraposition in sentential logic, which runs:

If 'p → q' then '¬q → ¬p'.

Applied to arguments formulated in ordinary language, this rule states: If you justify 'q' by means of 'p', the justification is consistent, and we wish to undermine 'q', then we will have to attack 'p'.

To apply this rule to textual interpretations we will need to briefly recapitulate what determines an interpretation. Characteristically, an interpretation is expected to provide answers to the following questions.

Demarcations: How is the text structured? How does the author separate a larger piece of argumentation from its context, and how are the

various individual arguments within that piece of argumentation separated from one another?

Emphasis: What emphasis is put on the various parts of the argumentation? What are the crucial sentences, and what sentences are less important? What expressions can we replace, or eliminate without replacement,
without compromising the consistency of the argumentation?

Concepts: What are the relevant concepts in the text? Do they deviate
from our contemporary terminology? Are there ambiguities?

Issues: What issues are discussed and examined by the interpreted
text, and how are these issues related to one another? What, if any, is
the main issue that the text is supposed to address?

These four items are closely connected. The issue that we determine
to be the text's primary focus has a direct impact on its structure, its
emphasis, and the relations between its relevant concepts. This also
applies vice versa, and in general, each of the four items above affects
every other item in the group. We often manage to grasp the specific
issue that the entire text is about and that thus provides the key to its
proper interpretation only late in the process of interpreting. This is
why it is best to start by demarcating the arguments, assigning emphasis in the text or with an analysis of the concepts.

The four items described above affect our interpretation of the original text in the following way. Demarcating the arguments in a certain
way, assigning certain levels of emphasis to the various sentences in the
text, analyzing the concepts in a certain way, and assigning the text
overall a certain topic or issue that it is primarily about will together
yield a particular interpretation, the result of which we can use as a
paraphrase of the text.

Applying the rule of contraposition to the currently predominant
interpretations gives rise to the following consideration:

> If demarcating the arguments, assigning emphases to different parts
> of the text, interpreting the concepts, and assigning the text an over-
> all topical issue the way it has been done in a particular interpretation
> together result in a completely absurd argument in Frege's text, then
> this absurd argument can be explained as a misunderstanding or
> misinterpretation only if we succeed in demarcating the arguments
> in a different way, putting different emphases on the various parts
> of the text, reinterpreting some concepts differently, or assigning the
> text a different topical issue than was done in that interpretation.

I shall demonstrate later, in the context of Frege's first and third arguments, how this rule is applied in a concrete case.

13.1.2   When comparing a given interpreter's paraphrase of an argument with the original text, we are often struck by an enormous distance between text and paraphrase, in which case we as readers do not quite know what emphasis the interpreter has placed on certain concepts, how he/she has demarcated the arguments from one another, and so on. This discrepancy can be considerably reduced if the interpreter provides abridged or otherwise modified quotations from the text. In this way, the reader is given the opportunity to directly notice which concepts were left out in the interpretation (because the interpreter did not regard them as relevant to the argument), how the interpreter has structured the text, and so on. As an interpreter I also indicated above which words I assign special importance to by presenting them in italics. By thus providing an *amended quotation*, the interpreter conveys his reading of the text to the reader; by contrast, the complete and unamended quotation would not tell us anything about the interpreter's reading. Every interpreter aiming at an interpretation that is easily verifiable should, therefore, quote the original text in a modified version as indicated above. (Of course, the modifications should be individually indicated, and the original text should be fully accessible to the reader in its unaltered form as well.)

Such a quotation, for example, lends itself to the following questions. What is the first and the last word of the quoted passage, that is, how – according to the interpreter – is it demarcated from other arguments? What expressions have been replaced by bracketed ellipses ('[...]'), indicating that the interpreter considers them irrelevant in the context of the argument? What expressions have been highlighted (through the use of italics or some other means), indicating that the interpreter assigns them special importance in the context of the argument?

In the following examples I apply the rule of contraposition to the omitted or deleted passages within the quotes: If a given interpretation involves the omission of certain expressions on the grounds that they are irrelevant in the context of the argument, what alternative interpretation do we obtain if we instead try to consider the omitted expressions as essential for the argument?

13.2   A road to the first argument

13.2.0   Frege's first argument is quoted by Künne as follows:

> (Z1) It might be supposed that truth consists in a correspondence.... Now a correspondence is a relation. But this goes against the use of the word 'true', which is not a relation word, does not contain any indication of anything else to which something is to correspond. (Künne, 1985, p. 123)

In this quote Künne modifies the text in two ways:

(1) The ellipsis replaces the words 'of a picture to what it depicts'.
(2) The quote as according to Künne starts out with 'It might be supposed'; Frege's original passage, however, starts out with 'It might be supposed *from this*'.

The words 'from this' connect the quoted passage to those that immediately precede it. Are these prior passages truly irrelevant to the first argument? Let us look at the passage in Frege's text that comes right before the one quoted by Künne:

> Obviously, we would not call a *picture* true unless there were an *intention* involved. A *picture is meant to* represent something. Neither is an *idea* called true in itself, but only with respect to an *intention* that it *should* correspond to something. (iii)

The fact that Frege refers to the bearers of correspondence truth, namely pictures or ideas, *four times*, and that he refers *four times* to the *intention* (2x 'intention', 1x 'should', and 1x 'is meant to'), is not reflected in Künne's quote. Now, Künne is perfectly justified in omitting these words in his quote *if* he considers them irrelevant to the argument at hand. And indeed, he does appear to consider the bearer and the intention to be irrelevant to the first argument, as his interpretation in 'Wahrheit' illustrates:

> (K1) Frege attacked [...] all versions of a correspondence theory: All of them allegedly fail on the grounds that 'x is true' is a one-place predicate while 'x corresponds to y', by contrast, is a two-place predicate. (Künne, 1985, p. 136)

This paraphrase of Frege's first argument refers neither to pictures or ideas nor to intentions. In Soames's paraphrase, moreover, the truth bearers and intentions are not assigned any importance either:

> The argument to this point is quite simple:
> 1. Grammatically, 'true' is a predicate applied to individual objects and so ought to stand for a property if it stands for anything at all.
> 2. Correspondence is a relation holding between at least two objects.
> 3. Thus truth is not correspondence – that is, the word 'true' does not stand for any correspondence relation (since it does not stand for any relation at all). (Soames, 1999, p. 24)

The same goes for Stuhlmann-Laeisz's paraphrase: 'A correspondence is a relation. The word "true", however, is a property word not designed to express relations at all' (Stuhlmann-Laeisz, 1995, p. 16).

We notice from Künne's, Soames's and Stuhlmann-Laeisz's paraphrases the same emphasis on portions of the text as we find in Künne's Frege quotes, and yet the quotes reveal this emphasis *at first sight*; hence it is much easier to notice by means of the quotes.

In short, all three interpreters agree that pictures/ideas as well as intentions are irrelevant to Frege's first argument. Under these circumstances, however, Künne would have been justified in truncating the quoted Frege passage even more than he did, as indeed he does in his later book *Conceptions of Truth*:

> (Z2) Now a correspondence is a relation. But this goes against the use of the word 'true', which is not a relation word.[1]

13.2.1   Ascribing to Frege the first argument in the way in which it is understood by Künne, Soames and Stuhlmann-Laeisz strikes me as completely absurd. But what can we do to prove that this interpretation is a misinterpretation? First of all, we should acknowledge that Künne's, Soames's and Stuhlmann-Laeisz's interpretation (hereafter shortened to 'I_{KS}') is internally consistent. If the first argument essentially relies on the trivial thesis that one-place predicates are not two-place predicates, then indeed it does not matter what truth bearers we select: the argument remains the same no matter whether we select pictures, ideas, sentences or thoughts as truth bearers. Hence it is legitimate to omit the truth bearers from the quoted passage in the context of this interpretation. We thus obtain the following implication:

(A) If interpretation I_{KS} holds, then the nature of the truth bearers is irrelevant to Frege's first argument.

The contraposition of (A) is:

(A') If the nature of the truth bearers is relevant in Frege's first argument, then interpretation I_{KS} does not hold.

Now, if we wish to show that I_{KS} is a misinterpretation due to its having completely absurd results, we should read the wording of the first argument all over again, more painstakingly and within a larger context comprising at least the first six paragraphs. In doing so we should ask ourselves whether the text might not be understood in a way in which the pictures and ideas are relevant components of the argument.

Along the way we will encounter the following question in the second sentence of the *third* paragraph:[2]

[(α)] [How] to delimit more closely the region within which truth could be predicated, *the region in which the question 'Is it true?' could be in principle applicable*[?] (i)

This question goes unanswered in the third paragraph, and is certainly not answered with the indefinability thesis at the end of the third paragraph. It is only at the end of the *fourth* paragraph that we find the answer to this question:

[(αα)] And hence the only thing for which the question 'Is it true?' can be in principle applicable is the sense of sentences. (ix)

Sentences (α) and (αα) bracket the two paragraphs, turning them into one argumentative unit concerning the nature of the truth bearers. Frege's most important argumentative goal in the third and fourth paragraphs consists in establishing the thought as the proper bearer of truth-values. This is why truth bearers are at the center of his argumentation in the third and fourth paragraphs, and the issue at the heart of the third and fourth paragraphs is: what is the proper bearer of truth – ideas/pictures, or sentences or thoughts?

Künne, Soames, and Stuhlmann-Laeisz linked Frege's first argument exclusively to the thesis at the end of the third paragraph that truth is indefinable, thereby neglecting to consider the fourth paragraph altogether in their interpretations of the first four arguments. The first argument contributes to *this thesis* according to I$_{KS}$: Truth cannot be defined as correspondence.

My considerations so far deviate from the previous interpretations in two respects: The *context* of Frege's arguments has been redefined and the central *issue* with which the text is concerned has been fundamentally reassessed. The discovery that the narrower context of Frege's critique of the correspondence theory comprises also the fourth paragraph, and that Frege's first argument can be linked to the question of the nature of truth bearers as well, already constitutes the beginning of a novel interpretation of the first argument since according to I$_{KS}$, the first argument does not contribute anything to the question of the nature of truth bearers. So let us try to look for an alternative interpretation according to which the first argument does contribute to answering that question.

13.2.2 In what way might pictures/ideas and intentions be relevant for the first argument? If I see a picture without knowing what it is supposed to represent, I do not have the knowledge I would need to determine whether or not the picture is true. Suppose that before me I see a picture of a large church with two towers. How can I determine whether this picture is an accurate ('true') image of Cologne Cathedral, which has two towers, or rather an inaccurate ('false') picture of the Ulm Minster, which has but one? Frege's criterion is the intention behind the picture:

> Obviously, we would not call a picture true unless there were an intention involved. A picture is meant to represent something. (iii)

The picture's truth-value depends on the intention with which it was painted; if it was painted without the intention of representing a particular object, then it has no truth-value whatsoever. Suppose that this picture is meant to represent Cologne Cathedral and does indeed depict that object accurately; in this case we can say:

> This picture corresponds to Cologne Cathedral, which it is supposed to depict; that is, it is true.

But is this how the word 'true' is used in the *sciences*, which is the very use on which Frege focuses? In Frege's view, if I have grasped a scientific thought such as the Pythagorean theorem, I do not have to ask anyone about the intention he or she associates with this thought. In particular,

I do not have to inquire about what Pythagoras's intentions were when he claimed that in any right triangle the area of the square whose side is the hypotenuse is equal to the sum of the areas of the squares whose sides are the two legs. The truth or falsity of scientific, mathematical, or historical thoughts such as '$a^2 + b^2 = c^2$' or 'The Eight Short Preludes and Fugues were not composed by J. S. Bach' is completely independent of any intentions that may be associated with such thoughts.

The comparison here is between the viewing of a picture and the grasping of a thought, or between a picture and a thought. While a picture can be true only relative to an intention, the truth of scientific thoughts is entirely independent of intentions. Frege is exclusively concerned with 'that sort of truth which it is the aim of science to discern'. Hence pictures and ideas, whose truth depends on intentions, do not qualify as bearers of scientific truth.

13.2.3   Though Frege applies the concept of truth as correspondence only to pictures and ideas, we can still ask whether the correspondence theory might not also be relevant with respect to sentences or thoughts as truth bearers. In the fourth paragraph it becomes clear that Frege applies the concept of truth primarily to thoughts and only secondarily to sentences. Frege uses the truth predicate especially in sentences such as 'It is true that it is raining' as opposed to sentences such as 'The sentence "It is raining" is true'.

It is certainly true that $a^2 + b^2 = c^2$, and likewise that the Eight Short Preludes and Fugues were not composed by J. S. Bach. But what should be the x to which the thought that $a^2 + b^2 = c^2$ corresponds, and to what y is supposed to correspond the thought that the Eight Short Preludes and Fugues were not composed by J. S. Bach? If the one relatum of correspondence is the thought then what is the second relatum? We could speculate that the second relatum is the corresponding state of affairs, but what is the difference between the thought that $a^2 + b^2 = c^2$ and the state of affairs that $a^2 + b^2 = c^2$? Translated into Frege's terminology, a state of affairs would be nothing but a thought, and an existing state of affairs would be nothing but a true thought. Thus there is no second relatum distinct from the thought – unless we decided to double reality in this respect.

The following point is crucial in all of these considerations: pictures are true only if they are associated with an intention. This intention provides the second relatum for the relation of correspondence. Thoughts, on the other hand, are true independently of intentions, and independently of whether they correspond to anything.

13.2.4 According to Künne, Soames, and Stuhlmann-Laeisz, the one-place predicates 'x is true', 'x is a wife', 'x is stolen', and so on, have one crucial thing in common: each of them can be defined by means of a two-place predicate. This is *precisely* why those interpreters consider Frege's first argument a fallacy. But is this view objectively justified? I would now like to draw the reader's attention to a crucial difference between these predicates.

The ordinary-language examples of one-place predicates offered by Künne ('x is a wife'), Soames ('x is a father'), and Stuhlmann-Laeisz ('x is married', 'x is stolen', 'x is rented out', 'x is usable'), can all be defined by two-place predicates in which one variable is bound by an existential quantifier. The qualifying two-place predicates have the following peculiarity. The sequence of words of the one-place predicate is contained in the sequence of words of the two-place predicate, as shown in table 13.1 below.

The predicates (a) – (f) express the same as the more explicitly formulated predicates (a') – (f'). Since, however, in (a') – (f') the particular person (or purpose) is not mentioned but merely referred to in an indeterminate way by means of 'somebody' (or 'something'), we can omit the expressions 'somebody's', 'to somebody', 'by somebody', and 'for something' for the sake of brevity; in fact, they are redundant. For we tend to answer the question 'Are you married?' with a short 'Yes', not with 'Yes, and to somebody'; the latter goes without saying.

Thus, the one-place predicate (a) 'x is a wife' can be defined by the two-place predicate (a') 'x is y's wife', ... and the one-place predicate (f) 'x is usable' can be defined by the two-place predicate (f') 'x is usable for y'. In general, for all ordinary-language examples of the kind offered by Künne, Soames, and Stuhlmann-Laeisz the following holds:

(A) 'x is P' can be defined by means of 'x is someone's P/P for/to/by someone'.

To be sure, 'x is P' does not have to be defined by means of 'x is someone's P/P for/to/by someone'; for at least some of the aforementioned one-place predicates can also be defined by means of other two-place predicates – for example, (a) 'x is a wife' by 'x is married to y', or (f) 'x is usable' by 'x serves the purpose of y'.

It is at least striking, however, that (A) holds of all the predicates (a) – (f), while it does not hold of 'x is true'. For if we follow Frege in applying the word 'true' to thoughts, then it cannot be defined

Table 13.1 One-place and two-place predicates

(a) 'x is a wife' =	(a') 'x is somebody's wife',
(b) 'x is a father' =	(b') 'x is somebody's father'
(c) 'x is married' =	(c') 'x is married to somebody',
(d) 'x is stolen' =	(d') 'x is stolen by somebody'
	[= 'x is stolen by somebody from somebody'],
(e) 'x is rented out' =	(e') 'x is rented out to somebody'
	[= 'x is rented out by somebody to somebody'],
(f) 'x is usable' =	(f') 'x is usable for something'
	[= 'x is usable by somebody for something'].

by means of 'x is true of y'. When we say 'It is true that it is raining here' we do not mean 'It is true *of something* that it is raining here', and if someone said: 'It is true of Bochum that it is raining here', it would just be but a bizarre way of saying 'It is true that it is raining in Bochum'. It is true that Cologne Cathedral is located right next to the Cologne central station, but of what is this true? Asking such a question means asking about pictures, not thoughts. If we were seriously committed to formulations such as: 'It is true of Cologne Cathedral that it is located right next to the Cologne central station', then we would need a different truth predicate: one that applies not to sentences or thoughts, but to predicates. We could then say that the predicate 'x is located right next to the Cologne central station' is true of Cologne Cathedral; or in other words, we could say, 'The predicate "x is located right next to the Cologne central station" can be truthfully asserted of Cologne Cathedral.' This might be a meaningful way of speaking. But it really does not say anything different from 'It is true that Cologne Cathedral is located right next to the Cologne central station.'

We can meaningfully say of a picture or a predicate that it is true of something, just as we can say of a wife that she is the (or a) wife of somebody. But we cannot say this of the complete thought that Cologne Cathedral is located right next to the Cologne central station; instead, we can only say that it is true, period. For in what sense could we say that this thought is true of Cologne Cathedral but false of the Bochum Town Hall or the moon? Suppose that Anna is a wife. We may ask whose wife she is. Suppose she is Peter's wife. Then this thought would be true. But the question of *of whom* or *of what* this thought is true is nonsensical. It is therefore false that the words 'true' (as used in science) and 'wife' can be *equally* construed as 'derivatives' (Stuhlmann-Laeisz) of two-place predicates, as has been claimed by Künne, Soames, and Stuhlmann-Laeisz.

The truth of pictures is a truth of something, which is why this kind of truth can be conceived as correspondence with something. But this use of 'true' goes against the use of the same word in scientific contexts, and therefore pictures cannot qualify as the bearers of scientific truth. From this point of view, the first argument assists us in determining the nature of the bearers of truth by ruling out pictures/ideas as bearers of scientific truth.

13.2.5   Let us return once more to the standard critique of Frege's first argument and try to understand how this misinterpretation could have come about. Like his earlier version in 'Wahrheit', Künne's critique of Frege's first argument in *Conceptions of Truth* focuses on the fact that the one-place truth predicate can in principle refer to a relational property and alleges that Frege denied this very fact. Künne writes:

> As an objection, Frege's argument is rather weak. To be sure, unlike 'agrees with' or 'corresponds to', the predicate 'is true' is a one-place predicate, hence it does not signify a *relation*. But the predicate 'x is a spouse' is also a one-place predicate, hence it does not signify a relation either, and yet it is correctly explained as 'There is somebody to whom x is married'. It signifies a *relational property*. [...] Perhaps the predicate 'x is true' can be similarly explained: 'There is something to which x corresponds'. If this explanation is correct, then, as Russell once put it, 'the difference between a true belief and a false belief is like that between a wife and a spinster'. (Künne, 2003, p. 94)

We can recapitulate more easily how Künne arrived at this critique if we take a closer look at his two quotations (see above) from Frege's first argument:

> (Z1) It might be supposed that truth consists in a correspondence [...] A correspondence is a relation. But this goes against the use of the word 'true', which is not a relation word, does not contain any indication of anything else to which something is to correspond.
>
> (Z2) A correspondence is a relation. But this goes against the use of the word 'true', which is not a relation word.

If we add the first sentence of (Z1) to (Z2) then we obtain:

> (Z1/2) [(a)] It might be supposed that truth consists in a correspondence [...] [(b)] A correspondence is a relation. [(c)] But this goes against the use of the word 'true', which is not a relation word.

The omission in the first sentence of this quote can easily create the impression that Frege here entertains the supposition that truth is a *relation* of correspondence. A short version of the argument contained in (Z1/2) looks like this:

(1) Truth consists in a correspondence. (according to (a))
(2) A correspondence is a relation. (according to (b))
(3) But 'true' is not a relation word. (according to (c))

However, Frege's original version of the first sentence is like this:

(β) It might be supposed from this that truth consists in a correspondence of a picture to what it depicts. (iii)

An elliptic version of this sentence would be:

(4) Truth consists in the correspondence of a picture with the thing depicted.

Sentence (4) makes it clear that truth is not to be conceived as the *relation* of correspondence, but as a *property*, namely the property of *correspondence to the depicted object*. Accordingly, the one-place predicate 'x is true' is not defined as having the same meaning as the two-place predicate 'x corresponds to y' but as having the same meaning as the one-place predicate 'Picture x corresponds to what it depicts', which does not refer to a relation but to a *relational property*, as we would call it today.

   Contrary to the standard (mis)interpretation and in accordance with the sentence (β) immediately preceding it, the next sentence in the quoted passage

(b) A correspondence is a relation

can be paraphrased as:

A correspondence [with something] is a relation [to something].

Because ordinary language rejects this sentence as redundant, we tend to leave out the two brackets, just as Frege did. The sentence immediately following (b), and containing the expression 'anything else' to indicate the second relatum, also supports the conclusion that Frege does not talk about correspondence as a relation but about the property of *corresponding to something*.

> But this goes against the use of the word 'true', which is not a relation word, [that is,] does not contain any *indication of anything else* to which something is to correspond. (iv)

Frege here explains what he means by a 'relation word' in this context:

> 'True' is *not* a relation word iff it does *not* contain any indication of anything else to which something is to correspond.

Hence,

> 'true' *is* a relation word iff it *does* contain an indication of something else to which something is to correspond.

In short: according to this quoted passage, a *relation word* of the kind that Frege has in mind here must contain an indication of something else. The *relation word* 'x corresponds to the object to which it is supposed to correspond' contains a general indication of the intended object, while 'x corresponds to Cologne Cathedral' contains a concrete indication of Cologne Cathedral.

There is, however, a second kind of 'relation word', which would be exemplified by expressions such as 'x corresponds to y' or simply 'correspondence'; such relation words do *not* contain any indication of something else. Let us distinguish these two meanings of 'relation word' as follows:

(a) Relation words such as 'correspondence' denoting actual relations and usually formulated as two-place predicates such as 'x corresponds to y'. Relation words in this sense do *not* contain any indication of something else. This meaning of 'relation word' conforms to contemporary linguistic practice.

(b) Relation words that are actually property words such as 'correspondence with the intended object'. Such relation words denote *relational properties* and are usually formulated as one-place predicates: 'x corresponds to the object to which it is supposed to correspond'. A relation word in this sense does contain an indication of something else. This meaning of 'relation word' is the one used by Frege in the quoted passage.

Künne, Soames, and Stuhlmann-Laeisz understood Frege's use of the phrase 'relation word' along the lines of (a), thus arriving at their own interpretation: 'A correspondence is a relation. But the word "true" is a property word [hence not a relation word] and not suited to express a relation' (Stuhlmann-Laeisz, 1995, p. 16).

But the passage quoted above makes it clear that in this context Frege regards a relation word as containing an indication of something else, and that he uses the expression 'relation word' along the lines of (b), contrary to what Künne, Soames, and Stuhlmann-Laeisz claim. Hence, in contemporary terminology we can paraphrase Frege's thesis in his first argument briefly as follows:

> The word 'true' in its scientific use is not a relation word of the kind described in (b), that is, it does *not* express a *relational property*.

One-place predicates, however, that do not refer to relational properties stand for absolute properties. In other words: Truth is an *absolute*, not a relational concept. With that we have arrived at the formulation that I used in Chapter 4 in my exposition of Frege's first argument.

13.3   A road to the third argument

13.3.0   While it is easy to follow Frege's first argument once we include the fourth paragraph in our interpretation and place pictures, ideas and intention at the center of focus, the third argument is still hard to follow even with that paragraph's inclusion. Here again, however, it is Künne's quotations from the third argument that enable us to understand it.

In the following I mark the passages that Künne slightly changed with bracketed letters. Künne's quote in 'Wahrheit' runs:

> [(a)] Could we not maintain that there is truth when there is correspondence in a certain respect? ... [(b)] And what would we then have to do so as to decide whether something were true? We should have to inquire whether it were true that, for example, an idea and something real correspond in the specified respect. And with that we should be confronted again by a question of the same kind, and the game could start all over. So this attempt to explain truth as correspondence breaks down. [This is immediately followed by a complete and unaltered quotation of the fourth argument.] (Künne, 1985, p. 137)

In (a) the quotation omits the question 'Or does it?' ['Oder doch?'], which provides a close link to the second argument; without that argument, we cannot grasp the question's motivation. There still is a faint reference to the previous text in the 'not' within the question 'Could we not maintain that [...]?' And in the place of (b) Frege's text has the question 'But in what respect?'

As we can see from Künne's summary paraphrase of the third and fourth arguments, he considers the 'respect' as irrelevant for an adequate understanding of the argument, which is why it is entirely omitted from the paraphrase:

> Suppose we define being true as being Φ. Then according to Frege's argument, on order to decide whether x is true we have to decide whether x is Φ. But to decide the latter means deciding whether it is true that x is Φ [...]. In order to decide this, the definition tells us to find out whether the thought that x is Φ is itself Φ, and so on ad infinitum. Thus, in order to find out whether x is true we would have to make infinitely many other decisions – which is impossible not only for medical reasons. (Künne, 1985, p. 137)

If the 'respect' indeed turns out to be irrelevant for the argument we may omit it in our quote, as Künne does in his *Conceptions of Truth*:

> [(c)]/[(d)] What would we... [(e)] have to do so as to decide whether something were true? We should have to inquire whether it were true that, for example, an idea and something real correspond... [(f)]. And with that we should be confronted again by a question of the same kind, and the game could start all over. So this attempt to explain truth as correspondence breaks down. (Künne, 2003, p. 129, modified standard translation)

In (c) and (d) we are missing the questions 'Could we not maintain that there is truth when there is correspondence in a certain *respect*?' and 'But in *what [respect]*?' In (e) we are missing 'and [...] *then*' (that is, 'after specifying the *respect*'), and finally in (f) we are missing 'in the specified *respect*'. All four text fragments (c) – (f) deal with the *respect* in which the correspondence is to hold. Thus, this short quotation entirely omits *four* references by Frege to the respect of correspondence! In addition, (c) represents the elimination of even the last faint reference to the previous argument (see the 'not' in the omitted sentence).

Table 13.2   The shortened third argument from 'Der Gedanke'

'Logik' (1897)	'Der Gedanke' (1918)
[...] decide [...] whether an idea did correspond to reality, in other words: whether it is true that the idea corresponds to reality.	[...] decide whether something were true? We should have to inquire whether it were true that, for example, an idea and something real correspond [...]

The result is a text that corresponds to the third argument in 'Logik', as shown by table 13.2.

With his omissions in the quoted text Künne thus reversed the very changes that Frege made in 'Der Gedanke' with regard to his earlier text from 'Logik': Künne entirely eliminated Frege's references to the *respect* and to the *preceding* arguments, which Frege had added to the text of 'Logik'. Accordingly, Künne reads the parallel passages in 'Logik' and in 'Der Gedanke' as expositions of the same argument (see section 7.9).

These omissions would be justified if the changes had obscured Frege's previous text or unnecessarily prolonged it with irrelevant material. But they are not justified if he, by contrast, improved his text by means of those additions. A charitable interpreter should assume that Frege did manage to improve his old text and that the changes are important for our understanding of the altered argument.

In 'Logik', the third argument is a special case of the fourth, and both arguments are most closely connected and marked off from their surrounding contexts. Both allege that the definitions of truth involve a circle: the third argument is directed against the correspondence theory, and the fourth against any definition of truth. Both arguments have only one exclusive argumentative goal, namely the indefinability of truth. In the same way Künne, as his paraphrase shows, understands the third and fourth arguments in 'Der Gedanke' as a unit, and on this basis it seems legitimate as well as sound to omit the 'respect' both in his quote and in his paraphrase of the argument; for under these assumptions it becomes irrelevant and even distracting to the argument thus interpreted. We thus obtain the following implication:

(B) If the third and fourth arguments in 'Der Gedanke' are as closely connected as the corresponding passages are in 'Logik' then the respect in which the correspondence is to hold is irrelevant for the third argument.

The contraposition then reads:

> (B') If the respect in which the correspondence is to hold is relevant for the third argument, then the third and fourth arguments in 'Der Gedanke' are not as closely connected as the corresponding passages in 'Logik'.

Now, if we regard the above regress argument as too absurd to have been concocted by Frege himself, we should try to find out in what way the respect in which the correspondence is to hold may be relevant to the third argument. My investigation of this question has prompted me to come to a very novel understanding of Frege's third argument.

First I noticed that though there is no meaningful link between the 'respect' and the fourth argument, there is a direct and close connection to the second argument. The second argument deals with correspondence in *every* respect, while the third argument deals with correspondence in a *certain* respect. Furthermore, Frege indicates this link between the two arguments by his question 'Or does it?' Thus, upon repeated reading of the text and much thinking, I have come to the conclusion that the four arguments in the third paragraph should be demarcated in a radically new way. The first three arguments concern I-truth and contribute to determining the nature of truth bearers, while the fourth argument is an excursus to establish the indefinability of truth. The third argument is thus much closer to the earlier two arguments than to the fourth.

Second, I attempted to place the question 'But in *what [respect]*?', which Künne omitted from the text, at the center of the third argument. The following is intended as a succinct recapitulation of this attempt (cf. Pardey, 2004, chapter 6).

13.3.1   Let me quote once more the text of the third argument while marking off the various questions by letters (a) – (e):

[(a)] Or does it? [= does truth admit of more or less?]

[(b)] Could we not maintain that there is truth when there is correspondence in a certain respect?

[(c)] But which respect? And what would we then have to do so as to decide

[(d)] whether something were true? We should have to inquire [(e)] whether it were true that, for example, an idea and something real correspond in the specified respect.

And with that we should be confronted again by a question of the same kind, and the game could start all over.

It is essential for our understanding of this argument to ascertain of which of the five questions (a) – (e) it is being said that it is followed by a question of the same kind (providing that it is indeed one of these five questions that Frege means when he says that after asking it we should be confronted again by a question of the same kind). Dummett, Künne, Soames, and Stuhlmann-Laeisz all claim that (d) and (e) are the two questions said to be of the same kind. But this view faces the problem that questions (d) and (e), if we interpret them à la Dummett and the rest, are really *one and the same* question rather than two questions of the same kind (see section 6.4). Since I now prefer to place the respect in which there is to be a correspondence at the center of the argument, I accordingly choose (c) as the question followed by a question of the same kind. Let us see whether this selection may help us to come to a different understanding of the argument as a whole. Though question (c) occurs only once in the quoted passage, it could objectively confront us again immediately following question (e).

Let me begin with an example. Suppose we are looking at a picture of Cologne Cathedral. Whether this picture is true or not depends on the respect in which the correspondence is to hold. Thus, the question arises,

(c'): In what respect is there a correspondence?

The respect in which the correspondence is to hold shall be determined by the following definition:

(D_C) x is true:= x corresponds to the intended object with respect to color (or shape).

This stipulation is an answer to question (c'). But what do we have to do, after stipulating the respect in which the correspondence is to hold, in order to determine the answer to the following question?

(d') Is the picture of Cologne Cathedral true?

According to Frege, we would need to first find the answer to the following question:

(e') Is it true that the picture corresponds to Cologne Cathedral with respect to its color (or shape)?

'With that', that is, following (e'), we are once again confronted with a question of the same kind as (c'), namely,

(c'') If the word 'true' as it occurs in (e') is to be understood also as a correspondence, then in what respect does *this* correspondence hold?

Certainly not with respect to color (or shape), as this has just been stipulated in our answer to (c')! For the concept of truth in (e') applies to the *thought* that the picture corresponds to the cathedral with respect to color, and thoughts are not colored objects (or objects that have a physical shape like those in the spatiotemporal world). This is why they cannot possibly correspond to something with respect to color (or shape). Thus, we cannot stipulate once and for all: 'Truth consists in a correspondence with respect to color (or shape)', as this would amount to nonsense with regard to (e'). Hence, we would need *one* concept of truth for our picture of Cologne Cathedral (correspondence with respect to color (or shape)), and *another one* for the thought in which the correspondence of the picture with Cologne Cathedral is represented.

Yet Frege presumably is looking for a *univocal* concept of truth, and in particular for one that satisfies the requirement that the two occurrences of 'true' in the question

(A) Is it true$_2$ that x is true$_1$?[3]

be synonymous. As the argumentation of (c') – (c'') has now shown, if truth$_1$ is defined as correspondence in a certain respect (for example, with respect to color (or shape)), then truth$_2$ has to be defined differently. Hence the concept of truth cannot be defined as correspondence in a certain respect – which is what was to be shown.

In other words: the fact that (c'') is a question of the same kind as (c') that has to be answered differently is inconsistent with the univocality of the concept of truth. Dummett, Künne, and the rest read the sentence 'And with that we should be confronted again by a question of the same kind, and the game could start all over' as being about an infinite regress. But the sentence is consistent with our being confronted twice with a question of the kind of (c) that has to be answered differently in each case.

13.3.2   Let me present this argument once more in a somewhat different way and with a little more distance from Frege's original text. Frege starts out with the following presumption:

(1) The two expressions 'true$_1$' and 'true$_2$' in question (A): 'Is it true$_2$ that x is true$_1$?' have the same sense.

The assumption now to be refuted is (2) below:

(2) Scientific truth can be defined as correspondence of pictures/ideas (with something else) in a certain respect R* – for example, as correspondence with respect to color (or shape).

(3) While Frege explicitly indicates the two relata for truth$_1$ (= correspondence) as 'for example, an idea and something real', he does not do so with regard to truth$_2$. Yet Frege's formulation of (e) makes it clear that he applies the predicate 'true$_2$' to a thought, namely by asking
     (e) whether it were true [= true$_2$] that, for example,
     an idea and something real correspond in the specified respect [= are true$_1$].

(4) Now if even truth$_2$ is to be conceived as a correspondence between a thought and something else x then what would be the respect in which they were to correspond? Certainly it cannot be a correspondence with respect to color (or shape).

(5) Hence the two expressions 'true$_1$' and 'true$_2$' in question
     (A) Is it true$_2$ that x is true$_1$?
     do not have the same sense. Thus assumption (2) is inconsistent with the trivial presumption (1) and therefore false. Hence pictures or ideas cannot be bearers of scientific truth.

Here again we see that in my interpretation the third argument is a contribution to the task of determining the nature of truth bearers, while in Dummett's, Künne's, and the others' interpretations this is not the case.

Unlike pictures or ideas, thoughts are very well suited to be the bearers of truth. For they pass the test of univocality, of the identity of $truth_1$ and $truth_2$. The two occurrences of the truth predicate in the question

(A$_0$) 'Is it $true_2$ that it is $true_1$ that it is raining here?'

both apply to thoughts, and hence they are synonymous. Thus, the above test does not undermine the thesis that the scientific concept of truth applies to thoughts.

In this way, placing the question of the respect in which the correspondence is to hold at the center of the third argument enables us to see easily that (e) is about two different truth concepts, which in my view is the crucial point of the third argument. Whether we emphasize the univocality of the concept of truth, as we just did here, or rather the independence of truth, as I did in Chapter 6, is another question that I do not need to address at this point. In either case, we see here again how applying the rule of contraposition to existing alternative interpretations has substantially aided us in understanding the argument under scrutiny. For these alternative interpretations overlook the importance of the respect in which the correspondence is to hold and accordingly omit the corresponding expressions in their quotes of Frege's texts.

13.3.3 Reading the third and fourth arguments in 'Der Gedanke' in accordance with their parallel passage in 'Logik' will support the appearance that both arguments essentially are one and the same, with the third being but a special case of the fourth. This demarcation, in which the third argument is closely linked to the fourth, has a disastrous effect on our interpretation of the two arguments. For assuming that my interpretation is right, the third and fourth arguments display the following characteristics, respectively. The third argument deals with *two different* concepts of truth, namely I-truth and S-truth. It shows how I-truth depends on S-truth in that the words 'It is true that' have to be inserted into the initial question 'Is idea I_0 true?' Only this *insertion* has to be performed here (replacement of the definiendum by the definiens is trivial). In the fourth argument, which is a circle argument, the very *same* concept of truth that is defined is also required for the application of the definiens. Thus, there is only *one* concept of truth, namely S-truth. The definition of truth is an analytic definition whose application calls for *decomposing* the initial question 'Is it true that...?' into subquestions.

Now if we consider both arguments as being essentially one and the same argument, then these characteristics cancel one another out: the third argument deals with *two* concepts of truth, but the fourth only with *one*. The fourth argument requires *decomposing* the initial question, but for the third it suffices that the words 'Is it true that' are *supplemented*. Accordingly, Dummett, Künne, and the rest overlooked that the third argument deals with two concepts of truth, because they saw only one concept of truth in the fourth argument. Likewise, they failed to see that the fourth argument requires decomposing, because supplementing the words 'Is it true that' suffices for the third. If the third and fourth arguments were essentially the same, such projections would be justified. But if they are essentially different arguments, those projections will inevitably result in misinterpretations. Thus, once again we see how important the demarcation of arguments is for their accurate interpretation.

13.3.4   To be sure, the road to a novel interpretation is not always as smooth as my description might make it appear. We frequently get trapped in dead ends, which I have omitted here, and need to look for a new road to continue on. But it should have become clear that we do not need to put the already existing interpretations aside, even if we consider their results to be inaccurate, but that – and how – we can use them constructively in the development of a novel interpretation.

14
Absolute or Relative Truth?

14.1 A generalized formulation of Frege's objections

14.1.0 Up to now, Frege's argumentation has turned out to be (internally) consistent; but is it still relevant today at all? Since Frege's objections to the correspondence theory originally have their place in the confrontation of I-truth and S-truth, we might think them to be no longer relevant. Scientific truth is now uncontroversially identified with S-truth; I-truth does not appear to play any part at all. This raises the possibility that Frege's objections to truth-as-correspondence are fatally infected by the obsolescence of the dispute within which they are originally framed – the dispute that, in fact, motivates them. If that is so, then eliciting the true nature of those objections by attending to their contexts secures Frege against the charge of obvious falsehood at the cost of consigning him to archaism – a questionable trade-off and, in any case, hardly a stirring defense!

Frege's objections, however, don't only undermine the correspondence theory of I-truth; they undermine any conceivable account of relative truth.

To see this, let us begin by recalling the most important results of Frege's argumentation in the third and fourth paragraphs. In 'Logik' Frege concludes his proof of the undefinability of truth with the following words:

> Truth is obviously something so primitive and simple that it is not possible to reduce it to anything still simpler. Consequently we have no alternative but to bring out the peculiarity of our predicate by comparing it with others. (Frege, 1897, p. 140 (129))

Table 14.1 Comparison of I-truth and Sense truth

I-truth:	Sense truth:
relative	absolute
imperfect	perfect
dependent	independent
definable	indefinable

It is not the case that a concept can only be either defined or unspecified and undistinguished from other concepts altogether. In 'Der Gedanke', too, Frege brings out 'the peculiarity of our predicate by comparing it with others' (see above). The object of comparison is here first of all I-truth. When I-truth and sense truth are compared, important differences are brought to light, as shown by table 14.1 above.

Absoluteness, perfection, and independence are not *characteristics* of truth; they do not belong to the content of truth. Rather, they are *properties* of truth – that is they are second-order properties. Grammatically this is indicated by the fact that they are used as adverbs: a thought is absolutely true, it is perfectly true, and it is independently true. Absoluteness, perfection, and independence are not properties of thoughts (in which case they would be first-order properties) but determine the manner in which a thought is true.

Another property of sense truth is its indefinability, which also clearly distinguishes it from I-truth. We should pay attention to two factors here: first, I-truth *is* definable but three complementary definitions of I-truth show that it is *not* proper, scientific truth. Secondly, sense truth is *not* definable but nonetheless *is* proper truth.

Furthermore, Frege determines the sense of a sentence as the proper bearer of truth. He first does so indirectly in the third paragraph by showing that the truth of ideas is not proper truth; it follows from this that ideas cannot be proper bearers of truth. Therefore, only the sentence and its sense remain as potential proper bearers of truth. In the fourth paragraph he conducts a direct demonstration that the sense is the proper bearer of truth by reducing I-truth to sentence truth and sentence truth to sense truth.

14.1.1   Now, with that as backdrop, consider the following reconstruction of Frege's argumentation. The reconstruction is a *re*-construction in that it generalizes Frege's claims and arguments so that they encompass, not merely I-truth, but any form of relative truth. I indicate such expansions here by adding 'I-truth' in parentheses, treating it as a special case of relative truth. In addition, I distinguish here between arguments (1) – (5) and conclusions (C₁) – (C₃). Furthermore, I add Frege's argumentation from the fourth paragraph (reduction of I-truth to S-truth, see (3b)) to the third argument.

> (1) Scientific truth is absolute, that is, not relative (and therefore does not consist in a correspondence),
>
> (2) unless the relata are identical. (The identity of the only relatum is compatible with the absoluteness and the perfection of truth.)
>
> (C1) Therefore, no relative (see (1)) truth with two different (see (2)) relata, and certainly not a correspondence truth (such as I-truth), can qualify as scientific truth.
>
> (C2) Therefore, the bearer of a relative truth (ideas as bearers of I-truth) cannot qualify as bearer of scientific truth.
>
> (3) (a) Relative truth (I-truth) is dependent on the truth of thoughts
>
> (b) and can be reduced to the latter. The truth of thoughts is absolute and not reducible to any other truth.
>
> (C3) The thought is the bearer of absolute truth.
>
> (4) The absolute truth of thoughts is indefinable because each attempt to apply a definition of truth fails due to a circle.
>
> (5) If we relativize the truth of thoughts, for example, by interpreting such truth as correspondence then we enter into an infinite regress when determining the truth bearer.

Thus considered, Frege's argumentation can be seen as a critical examination of the opposition between *absolute* and *relative* truth. Frege distinguishes absolute truth from relative truth and defends the former against all claims of the latter. The identification of the thought as the proper bearer of truth is also meant to serve this aim. Frege's conception of truth can be briefly summarized in one short sentence:

Truth is absolute, that is, not relative, and its bearer is the thought.

The generalized arguments are so similar to Frege's original arguments that they can be derived from the latter by some minor and unproblematic modifications.

14.2   Absolute truth of thoughts versus relative truth of predication

14.2.0   According to Frege, the absoluteness of truth is incompatible with the correspondence theory of truth because the latter treats truth as relative. Now, there are proponents of a correspondence theory who endorse both the traditional correspondence theory ('veritas est adaequatio rei et intellectus') and the absoluteness of truth, but do not see any contradiction between the two. Why should there be a contradiction, and why should Frege be the first to discover this contradiction given that the correspondence theory has been around for so many centuries? This objection is due to Markus Stepanians during a discussion of my interpretation of the first argument.

In responding to this objection, I begin by distinguishing between traditional, Aristotelian logic as a logic of *predication* and modern, Fregean logic as a logic of *thoughts*. (This distinction is intended only as a very rough account.) Predication is relatively true while thoughts are absolutely true. Both concepts of truth are in a similar relation to one another as I-truth and S-truth are in Frege. I shall now proceed to argue for this relation in more detail.

An utterance of the sentence 'Socrates is human' can be analyzed in different ways. Frege distinguishes between the thought expressed by the *entire* sentence and the assertion. In Aristotelian logic, however, the focus is on the subject–predicate structure of the sentence, and the sentence is structured accordingly: of Socrates, the *subject* of the sentence, is *predicated* the property of being human; the assertion is here contained in the predicate. Socrates is named by means of the name 'Socrates', but this *naming* cannot be true or false; truth and falsity can be possessed only by the *predication* (or denial) of properties. Accordingly, if we say 'That is true' in response to an utterance of the sentence 'Socrates is human' we could mean either of two things. We could, following Frege, be applying truth to the *sense* of the *entire* sentence. Alternatively, we could, in the vein of traditional logic, be applying truth to the *predication*, that is, to only a *part* of the sentence: have we predicated (the property of) being *human* truthfully of Socrates? Does this predicate apply to Socrates? Whether the predicate applies depends on of whom it is predicated. If the property of being human is predicated of Socrates then the predication is true; if it is predicated of an animal or a rock then it is false. That is why the truth of predication is always a form of relative truth:

The property of being human is predicated with truth of Socrates.

Frege himself would hold that this last sentence contains not only the relative truth of predication but also the absolute truth of a thought. For due to the omnipresence of absolute truth this sentence is synonymous with the following sentence:

It is (absolutely) true that the property of being human is predicated with (relative) truth of Socrates.

Thus we can show with Frege that the relative truth of predication depends on the absolute truth of thoughts and can be reduced to the latter. The predicate is the bearer of relative truth while the thought is the bearer of absolute truth, and in this way both concepts of truth can be reconciled.[1] This, or a similar stance, may well have been taken by logicians or philosophers prior to Frege.

Thus we can say: a predication of an object is true if and only if the predication corresponds to the object. The correspondence theory thus emerges as an adequate theory of the relative truth of predication, but that does not mean that it is an adequate theory of the absolute truth of thoughts. We can endorse a correspondence theory for the relative truth of predication *and* at the same time maintain the absoluteness of the truth of thoughts without contradicting either ourselves or Frege's position. A contradiction with Frege's position would arise only if we promoted the relative truth of predication to the status of proper, scientific truth.

14.2.1 The transition from a logic of predication to a logic of thoughts is not without consequences for the concept of truth. And this is not surprising if we consider the (otherwise so often-cited) radical differences between Aristotelian and Fregean logic. In the new logic of thought we can no longer conceive of truth in terms of correspondence of predication with an object. Nor does it suffice to try to minimize the consequences by replacing the correspondence of a predication with an object by a correspondence of a thought with a fact. Even the thus-altered concept of truth as correspondence remains inconsistent with both the absoluteness and the perfection of scientific truth.

As paradoxical as this may sound, to my knowledge Frege is the only (critical) correspondence theorist who has thus far dared to accept the inevitable implications of modern logic by identifying facts with true thoughts: 'A fact is a thought that is true' (Frege, 1918a, p. 74). For if true thoughts correspond to facts then the former must be identical to the latter because scientific truth is absolute and perfect. From this point of view Frege's critique of the correspondence theory is that theory's dialectical sublation, in the Hegelian double sense of the word (supersession and preservation), under the conditions of modern logic.

14.3 Another objection

After Stepanians's critique I should like to discuss yet another objection to my interpretation. An anonymous reviewer of an earlier version of this book, whom I will call reviewer A, asked me to consider the following with respect to Frege's critique of the correspondence theory:

> There is, of course, a long and strong history of correspondence. The belief that there are truth-makers, for example, states of affairs, still appeals to some (D. Armstrong). Frege rejects this view (Facts are true thoughts!). Now, Frege held that the Bedeutung of a declarative sentence is related to the Bedeutung of its parts, and he identified the Bedeutung of a sentence with its truth value. Why isn't this correspondence?

There certainly is a relation between the *Bedeutung* (reference) of a sentence (its truth-value) and the *Bedeutung* of its parts, especially the *Bedeutung* of the proper names (the designated objects). But what kind of relation is it? If this relation is a correspondence – is it to be understood in the sense of the correspondence theory of truth? According to Frege the truth-value (the *Bedeutung*) of a sentence remains *unchanged* when a part of it is substituted by another expression with the same *Bedeutung* (see Frege, 1892a, p. 35). This connection can be described as follows:

> (N) If the names 'a'and 'b' have the same *Bedeutung*, that is, designate the same object, then the sentences 'F(a)' and 'F(b)' have the same *Bedeutung*, that is, the same truth-values.

> (P) If the predicates 'F(x)' and 'G(x)' have the same *Bedeutung*, the sentences 'F(a)' and 'G(a)' have the same *Bedeutung* (see Frege, 1892b, pp. 128ff (118ff)).

From (N) and (P) we obtain the following corollary:

> (NP) The *Bedeutung* of 'F(a)' (perfectly) coincides with that of 'G(b)' – that is, the two sentences have the same truth-value, – if the names 'a' and 'b' have the same *Bedeutung* and the predicates 'F(x)' and 'G(x)' also have the same *Bedeutung*.

But this is not the kind of correspondence that is required for a correspondence theory definition of truth, for (NP) merely asserts under what conditions the truth-value of 'F(a)' remains *unchanged*. (NP) does not assert, however, under what conditions the sentence 'F(a)' has the truth-value True (or False); that is, (NP) does not specify *which* truth-value the sentence has. By contrast, a definition of truth based on the correspondence theory would be expected to do just that: to specify which truth-value a sentence has under certain conditions. Hence such a definition would have the following consequence with respect to our schematic sentence above:

The sentence 'F(a)' is true if and only if it corresponds to its relatum A (however this relatum is to be determined) .

It would be different in the following situation. Let us specify the *Bedeutung* of the predicate 'F(x)' as the extension of this predicate, that is, as the set $\{x/F(x)\}$. Then let us define:

'F(a)' is true:= The *Bedeutung* of 'a' is an element of the *Bedeutung* of 'F(x)', that is, of the set $\{x/F(x)\}$.

With that we would move in the direction of Tarski's definition of truth. But this is not the connection described in (NP). Therefore, I do not see how we could ascribe a correspondence theory of truth to Frege on the basis of (NP).

14.4 The liar as the key witness against absolute truth

14.4.0   The reception of Frege's work was compromised by the discovery of antinomies on two separate occasions. Frege's foundation of logic and mathematics became known to the wider philosophical public only after and through the discovery of Russell's antinomy, while Frege's theory of truth began to be examined only after Tarski's analysis of the liar antinomy and his definition of truth had become generally established in the field of analytic philosophy. Thus, both Frege's logic and his semantics first came to the awareness of the philosophical public as theories that had already been *refuted*.

Tarski's analysis of the liar antinomy is (among other things) a plea for relativizing the concept of truth. Tarski's analysis quickly became common knowledge, rendering the concept of absolute truth obsolete even before Frege's theory of truth was acknowledged in the first place.

14.4.1   It is easy to see how to construct the liar antinomy with the help of Frege's absolute concept of truth. Let us look at the thought expressed in the following line:

(*) The thought in line (*) is not true.[2]

From the omnipresence of truth

(F) The sentences 'p' and 'The thought that p is true' are synonymous.

we can immediately derive a Fregean variant of Tarski's convention (T):

(T_F) The thought that p is true if and only if p.

If we apply (T_F) to (*) we obtain:

(a) The thought that the thought in line (*) is not true is true if and only if the thought in line (*) is not true.

Looking at line (*) we recognize the following identity:

(b) The thought that the thought in line (*) is not true = the thought in line (*).

Based on this identity, the names 'The thought that the thought in line (*) is not true' and 'the thought in line (*)' refer to the same thought; they thus can be substituted for each other (in extensional contexts). If we now replace the longer name in (a) by the shorter name, we obtain the following contradiction:

(c) The thought in line (*) is true if and only if the thought in line (*) is not true.

This is the liar antinomy in a version that involves thoughts as truth bearers and utilizes Frege's concept of absolute truth.

14.4.2   Tarski's solution to the liar antinomy relies upon the distinction between object- and metalanguage and the construction of a hierarchy of semantically unclosed languages (with the resulting relativization of the concept of truth). With those in place, the truth predicate may be used only in the form 'The sentence x is true in language S_n', where 'S_n' (for $n = 0, 1, 2, ...$) refers to the various levels of languages within the Tarski hierarchy. To avoid the antinomy it is crucial to follow the rule that the predicate 'Sentence x is true in language S_n' may occur only in sentences of *higher-level* languages $S_{n+k\ (k\ =\ 1,\ 2,...)}$. Thus, a sentence in a language of level S_n can be ascribed the property of being true in S_n only in a language of a higher language level S_{n+k}. In short: ascription of relative truth is performed always from top to bottom within the hierarchy.

I want to illustrate this by means of some examples. Let the object language S_0, which must not contain any truth predicate, contain the sentence

(a) Cologne Cathedral has two towers.

Then we can formulate in metalanguage S_1 the sentence

(b) The sentence 'Cologne Cathedral has two towers' is true in S_0

and subsequently in metametalanguage S_2 the sentence

(c) The sentence 'The sentence "Cologne Cathedral has two towers" is true in S_0' is true in S_1.

Each of the three sentences is assigned a truth-value on a certain language level: (a) is true in S_0, (b) is true in S_1, and (c) is true in S_2. None of the sentences is absolutely true, that is, none of them is true independent of a particular language-level. Tarski no longer recognizes the one absolute predicate 'x is true' but only the relative predicates 'x is true in $S_{n\ (n\ =\ 0,\ 1,...)}$'.[3]

14.4.3 Within Tarski's hierarchy of languages, the liar antinomy can no longer be derived. Instead, the rule is observed that a truth predicate 'true in S_n' can only occur in a sentence of a higher-level language $S_{n+k\ (k\ =\ 1,\ 2,...)}$, but not in a sentence of S_n itself or in a sentence of a lower-level language. A sentence violating this rule does not belong to any language within the hierarchy and therefore counts as meaningless within the hierarchy.

Let us now consider the sentence

(**) The sentence in line (**) is not true in S_n.

This sentence applies the truth predicate 'true in S_n' to the sentence itself, since it is the very sentence in line (**). But with that, the sentence violates the aforementioned rule and hence becomes meaningless in all languages of Tarski's hierarchy. Therefore, we can no longer construct the liar antinomy within Tarski's hierarchy of languages.

14.4.4   The result seems clear: Frege's absolute truth can be used to generate the liar antinomy, while Tarski's relative truth concept avoids it. That is why the concept of absolute truth is, at the very least, threatened by the liar antinomy, and perhaps Frege's thesis that truth is absolute is even refuted by the liar antinomy. In any case, under the circumstances it seems reasonable to give up the concept of absolute truth and to adopt instead Tarski's relative truth concept. Quine writes in connection with Tarski's investigations: '[...] perhaps a time will come when truth locutions without implicit subscripts, or like safeguards, will really sound as nonsensical as the antinomies show them to be' (Quine, 1966, p. 9).[4]

Frege's concept of absolute truth is a truth concept 'without implicit subscripts' and hence meaningless, as the liar antinomy has shown according to Tarski and Quine. What's more a meaningless concept, it seems, should be eliminated as soon as possible from any systematic philosophical vocabulary. This explains how in Tarski's shadow absolute truth, the major object of Frege's focus, could become entirely invisible.

14.4.5   My defense of Frege's argumentation would be incomplete if it passed over the liar's threat to absolute truth. For my main thesis is that Frege's argumentation can be reconstructed as an internally consistent piece of argumentation in virtue of Frege's distinction between absolute and relative truth. But if the concept of absolute truth were self-contradictory or meaningless then the distinction between absolute and relative truth would become obsolete as well. Though Frege's argumentation is no longer exposed to the objections offered by Dummett and other scholars, as I have shown, it would still fall victim to the self-contradictory or meaningless nature of the concept of absolute truth. The result would be the same: in the end Frege's argumentation would turn out to be a collection of mistakes.

I want to counter this common attack of absolute truth and to show in the next chapter that this argumentation does not at all force us to relativize truth.

15
Tarski's Definition of Truth and Frege's Critique

15.1 Frege's absolute truth and Tarski's hierarchy of languages

15.1.0 The languages in Tarski's hierarchy avoid the liar, and so the liar no longer poses a threat to Tarski's relative truth predicates 'x is true in language $S_{n (n = 0, 1, ...)}$'. But how can Frege manage to avoid the liar if he sticks to his absolute concept of truth? Doesn't Frege also have to relativize the concept of truth in order to avoid the liar? Building on considerations actually present in Frege's texts, I want to show in the following that and how this apparent necessity of relativizing absolute truth can be escaped.

Frege discusses the claim that the concept of truth may have to be relativized in several places. See for the following section 11.4, which includes lengthier quotes from Frege containing the following abbreviated quotes. In 'Der Gedanke' he writes:

> [It frequently occurs that] the mere wording, which can be made permanent by writing [...], does not suffice for the *expression* of the thought. [...] If a time-indication is conveyed by the present tense one must know when the sentence was uttered in order to grasp the thought correctly. Therefore the time of utterance is part of the *expression* of the thought. If someone wants to say today what he expressed yesterday using the word 'today', he will replace this word with 'yesterday'. Although the thought is the same its verbal *expression* [the sentence] must be different in order that the change of sense which would otherwise be effected by the differing times of utterance may be cancelled out. The case is similar with words like 'here' and 'there'. In all such cases the mere wording, as it can be preserved

in writing, is not the complete *expression* of the thought; the knowledge of certain conditions accompanying the utterance, which are used as means of *expressing* the thought, is needed for us to grasp the thought correctly. (Frege, 1918a, p. 64; my emphases, modified translation)

Much earlier, in his *Grundgesetze der Arithmetik*, he addresses the issue in a similar manner:

The sense of the word 'true' could not be more wickedly falsified than by incorporating a relation to those who judge! But surely, it will be objected, the sentence 'I am hungry' can be true for one person and false for another. The sentence certainly, but not the thought, since the word 'I' in the mouth of the other refers to [...] a different person, and hence the sentence uttered by the other expresses [!] a different thought. All specifications of place, time, and so on, belong to the thought whose truth is at issue; *being true* itself is placeless and timeless. (Frege, 1893, p. xvif)

15.1.1   Frege refuses to apply the concept of truth directly to the sentence 'I am hungry' as the proper truth bearer and to thereby relativize the concept of truth. For this would allow for the following possibility:

The sentence 'I am hungry' is true in the mouth of one person and false in the mouth of another. (Cf.: The sentence 'A billion is a thousand millions' is true in contemporary American English but false in (the older) British English.)

His strategy to avoid relativizing truth can be summarized in two steps. First, we apply the concept of truth to thoughts. Thus we obtain:

The thought expressed by the sentence 'I am hungry' is true when the one person utters the sentence and false when the other person does so.

This step by itself could still leave the truth of thoughts relative, for in the one situation the thought would be true and in the other it would be false. But Frege objects to this by emphasizing that the two situations do not share *one and the same* thought. Rather, depending on the situation of utterance the sentence will express a different thought. Instead of relativizing truth we can relativize the relation of *expressing*;

that is the second step. Accordingly, instead of the *two*-place predicate 'Sentence x expresses the thought that p' we now have a *three*-place predicate 'Sentence x expresses the thought that p in the mouth of person Z.' With that we can construe the following perfectly meaningful sentence:

The thought expressed by the sentence 'I am hungry' in the mouth of this person is *true*, while the (different!) thought expressed by the sentence 'I am hungry' in the mouth of that person is *false*.

Thus, by relativizing the relation of *expressing* to a specific speaker we can escape the alleged necessity of relativizing truth.

15.1.2 Tarski's relativization of truth to various language levels can be circumvented in a fully analogous manner. For this we need the following three-place predicate: 'Sentence x expresses the thought that p in language S_n.' If we apply this scheme to the sentences above we obtain the following paraphrases:

(a') Cologne Cathedral has two towers.
(b') The thought expressed by the sentence 'Cologne Cathedral has two towers' in language S_0 is true.
(c') The thought expressed by the sentence 'The thought expressed by the sentence "Cologne Cathedral has two towers" in language S_0 is true' in language S_1 is true.

15.1.3 We could construct a 'Fregean' hierarchy of languages H_F in analogy to Tarski's hierarchy H_T. Such a hierarchy could be roughly characterized as follows. Tarski's sentence scheme

Sentence 'p' is true in S_n

becomes, in the Fregean hierarchy,

The thought expressed by sentence 'p' in S_n is true.

If sentence 'p' is true in S_n, according to Tarski, then it is a sentence of S_n, thus – in Frege's terminology – 'p' expresses in S_n a thought. Hence each sentence 'p' that is true in S_n corresponds to a true thought expressed by sentence 'p' in S_n. This correspondence between the two hierarchies holds not only with respect to sentences and their quotation names, but also with respect to any labeled sentences, as the following example illustrates. The Tarski sentence

Pythagoras' theorem is true in S_n

can be translated into the Frege sentence:

The thought expressed by Pythagoras' theorem in S_n is true.

Hence, in general we can translate Tarski's predicate

(P_T) Sentence X is true in S_n

into the Fregean predicate

(P_F) The thought expressed by sentence X in S_n is true.

15.1.4   The liar antinomy cannot be constructed within the Fregean hierarchy of languages H_F either. For the languages within this hierarchy follow the rule that the predicate

(P_F) The thought expressed by sentence X in S_n is true

may occur only in a sentence of a higher-level language $S_{n+k\,(k\,=\,1,\,2,\,\dots)}$, but not in a sentence belonging to S_n itself or in a sentence of a lower-level language. A sentence that violates this rule does not belong to any language in the hierarchy and therefore counts as meaningless within the hierarchy. This excludes the possibility that a sentence within this hierarchy of languages ascribes or denies truth with regard to the thought expressed by itself. Consider the sentence

(#) The thought expressed by the sentence in line (#) in language S_n is not true.

This sentence applies the predicate (P_F) 'The thought expressed by sentence X in S_n is true' to itself, since it is the very sentence in line (#), and hence to a sentence of the same language to which it also belongs. With that the sentence violates the aforementioned rule and becomes meaningless in any language within this hierarchy. Therefore, the liar antinomy can no longer be constructed in the Fregean hierarchy. Yet the concept of truth within this hierarchy is still *absolute*! Thus the liar antinomy can be circumvented even with an absolute concept of truth.

According to the standard Tarski reception, if we want to escape the liar antinomy then we have to introduce a hierarchy of languages and, as a *consequence*, relativize the concept of truth. Yet in fact, as I have just shown, to escape the liar it suffices to introduce a hierarchy of languages; we do not have to relativize the concept of truth in addition. The introduction of a hierarchy of languages forces us to relativize truth predicates to *language levels* ('true in S_n') only if we have already relativized the truth concept to *languages* ('true in S'). The former type of relativization of truth necessarily follows from Tarski's selection of the *sentence* as truth bearer. If, however, we follow Frege in choosing the *thought* as truth bearer, we are in no way forced to relativize the concept of truth after introducing a hierarchy of languages. Thus, here again the choice of the truth bearer is decisive.

15.2   The impact of Tarski's recursive definition of truth

15.2.0   Up to this point I have shown that Frege's critique of the correspondence theory is internally consistent, and that the liar antinomy can be avoided even with an absolute concept of truth. However, Tarski's recursive definition of truth has been just as influential as his analysis of the liar antinomy and has been regarded as a counterexample against Frege's indefinability thesis. Therefore, I will now briefly address Tarski's definition of truth in the light of Frege's critique of the correspondence theory.

Tarski's definition of truth, to which Frege would probably not have any *immanent* objections (such as a circle objection, a regress argument and so on), substantially contributed to the fact that for a long time, Frege's critique of the correspondence theory – and especially his thesis that truth is indefinable – were generally believed to have been refuted. The following passage from Günther Patzig is representative of this kind of Frege reception:

[It appears that Frege's] thesis that the concept of 'truth' is indefinable [cannot be maintained.] [...] [For as] thorough investigations by Tarski and others have shown in the meantime, the concept of truth [...] can be defined on the basis of a distinction between object- and metalanguage [...]. (Patzig, 1966, pp. 22f)

But where precisely is the contradiction between Tarski's definition and Frege's indefinability thesis?

15.2.1   To answer this question I shall first use an example to illustrate what a *recursive* definition of truth is. Let us consider the language G*, which despite being rather basic is entirely sufficient for our purposes. (I have adopted this example from Künne, 2003, pp. 192ff.) Language G*, which consists of a tiny fragment of German, can be specified by indicating the vocabulary, syntax, and semantics for that language.

The *vocabulary* of G* consists of the two elementary sentences 'Die Erde bewegt sich' and 'Der Mond ist rund', the sentential operators 'und' and 'Es ist nicht der Fall, dass...', and the two parentheses '(' and ')'.

15.2.2   The *syntax* of G* lays down by way of a recursive definition what is to be regarded as a sentence of language G*.

For all expressions S: S is a *sentence in G*:=*
(a) S = 'Die Erde bewegt sich', or

(b) S = 'Der Mond ist rund', or

(c) S = ⌜(S_1 und S_2)⌝, and S_1 and S_2 are sentences in G*, or

(d) S = ⌜(Es ist nicht der Fall, dass S_1)⌝, and S_1 is a sentence in G*.

The symbols '⌜' and '⌝' are quasi-quotation marks. ⌜(S_1 und S_2)⌝ is that sentence in G* which we obtain by first writing the parenthesis '(', then adding the sentence S_1, then the German expression 'und', then the sentence S_2 and finally the parenthesis ')' in the indicated order. ⌜(Es ist nicht der Fall, dass S_1)⌝ is that sentence in G* which we obtain by first writing the parenthesis '(', then adding the German expression 'Es ist nicht der Fall, dass', then the sentence S_1 and finally the parenthesis ')' in the indicated order.

15.2.3   Similarly, the *semantics* of G* gives us a recursive definition of 'true in G*'.

For all sentences S: S is *true in G*:=*
(e) S = 'Die Erde bewegt sich', and the earth moves, or

(f) S = 'Der Mond ist rund', and the moon is round, or

(g) S = ⌜(S_1 und S_2)⌝, and both sentences are true in G*, or

(h) S = ⌜(Es ist nicht der Fall, dass S_1)⌝ and it is not the case that S_1 is true in G*.

15.2.4   The following example illustrates how to determine the truth-value of a particular sentence by means of the above rules. According to rule (h) the sentence

(1) Es ist nicht der Fall, dass (die Erde bewegt sich und der Mond ist rund)

is true in G* if and only if the sentence

(2) Die Erde bewegt sich und der Mond ist rund

is not true in G*. According to rule (g) this sentence is not true in G* if and only if at least one of the following two sentences is not true in G*:

(3) Die Erde bewegt sich.

(4) Der Mond ist rund.

According to rule (e) sentence (3) is not true in G* if and only if the earth does not move, and according to rule (f) sentence (4) is not true in G* if and only if the moon is not round. Thus, sentence (1) in its entirety is true in G* if and only if the earth does not move or the moon is not round.

15.3   Tarski's recursive definition versus Frege's analytic definition

15.3.0   In the recursive definition of 'true in G*', the truth-value of a complex sentence is specified by means of the truth-values of its less complex components in accordance with rules (g) and (h), and the truth-values of the basic sentences are specified according to rules (e) and (f). To be sure, the recursive definition will not tell us what the *sense* of the expression 'true in G*' is. Yet, even without our knowing the sense of 'true in G*', the rules enable us to determine that sentence (1) is true in G* if and only if the earth does not move or the moon is not round.

Though the definiendum 'true in G*' also occurs in the definiens, namely in rules (g) and (h), the definition is nonetheless immune to circle objections, since rules (e) and (f) contain the conditions under which the *basic* sentences are true, and we do *not* need to know the sense of the predicate 'true in G*' in order to understand these conditions. When it comes to *complex* sentences, moreover, the truth conditions can be reduced to the truth conditions of the basic sentences, so that in general, we can specify the truth conditions of any complex sentences in G* based on the above rules without prior knowledge of the sense of the predicate 'true in G*'.

Thus, on one hand, the definition is non-circular, and on the other hand, it does not specify for us the sense (or content) of the predicate 'true in G*'. In this respect a *recursive* definition such as the one Tarski has presented to us is not an *analytic* definition in Frege's sense. A recursive definition serves to specify the extension of a predicate without requiring any knowledge or information about the content of the same predicate. An analytic definition, by contrast, is used to specify the content of a predicate. Thus, we can say that Tarski's recursive definition is a definition of an *extension*, while Frege's critique is directed against a definition of *content*, that is, a definition that analyzes the content of a predicate.

15.3.1   The relation between the content and the extension of a concept can be characterized as follows: although by specifying the content of a concept we thereby also specify its extension, the reverse is not the case, since several distinct conceptual contents may correspond to one and the same extension. A classic example is the pair 'rational creature' and 'featherless biped', which have different contents yet the same extension – namely the set of all humans (leaving aside plucked chickens and the like). If Frege's claim were that there is no definition of the extension of truth and Tarski's were that he can present a definition of the content of 'true', then, though referring to different kinds of definitions, they would indeed contradict each other. For a definition of a conceptual content is at the same time a definition of extension, because specifying the content automatically specifies the extension of a concept.

Yet in the present case the situation is quite the opposite: Frege denies the possibility of a(n analytic) definition of content, while Tarski proposes a (recursive) definition of extension, and these two positions are entirely consistent with one another. Thus, the alleged contradiction between Tarski's definition of truth and Frege's indefinability thesis is at *this* point revealed to be mere appearance, for there is no contradiction in the claim: though there can be no *analytic* definition of the concept of truth, as Frege has shown, there is still the possibility of a *recursive* definition, as Tarski has demonstrated.

15.4   Frege's critique and Tarski's definition of truth

15.4.0   The comparability of Frege's critique of the correspondence theory and Tarski's definition of truth is limited. Is Tarski's theory of truth really a correspondence theory, as has been claimed by Popper, Davidson, and Tarski himself? If we look at the above recursive definition of truth, or even more complex ones, we will have a hard time accepting this claim. Künne has given a negative answer to the question 'Does Tarski hold a *correspondence theory* of truth?' and convincingly refuted the various justifications offered by Tarski, Popper, and Davidson in favor of the opposite view (Künne, 1985, pp. 157ff, and 2003, pp. 208–225).

   If Künne is correct, Tarski does not in fact hold a correspondence theory. As a result, Frege's critique of the correspondence theory would not apply to Tarski's definition of truth at all. Conversely, it would be inappropriate to say that Frege's critique of the correspondence theory was refuted by Tarski's definition of truth, since the theory proposed by Tarski is not a correspondence theory in the first place – which is why the two conceptions of truth are not incompatible at this point.

15.4.1   Frege's first and second arguments directly target his definition of I-truth as *correspondence*. Since Tarski's formal definition does not involve correspondence, Frege's objections cannot be directly applied to Tarski's definition. But Frege assumes *one single* absolute and perfect concept of truth, which would rule out his acceptance of Tarski's replacement of this *one single* truth concept by a family of *distinct* truth concepts 'X is true in S_0', 'X is true in S_1', ... because it would mean a relativization of truth. In any case, Tarski does not define Frege's absolute truth predicate 'x is true', but a multitude of other truth predicates. Thus Tarski is far from proposing an alternative conception of absolute truth; instead, he declares Frege's truth concept to be meaningless and replaces it by a multitude of non-absolute truth concepts. From Frege's point of view, however, we should comment on this replacement as follows. The absolute, scientific concept of truth is not covered by Tarski's definition. For this reason, Frege would regard Tarski's definition of 'x is true in $S_{n (n = 0, 1,...)}$' as being no less inadequate than the definition of I-truth; neither definition qualifies as a definition of absolute truth. Furthermore, Frege rejects the *sentence*, Tarski's truth bearer, as bearer of absolute truth, because the sentence, no less than the idea, can only be relatively true.

Thus Frege and Tarski differ not only in their respective views of definitions but also and especially in their accounts of *truth*. If we say, without further explanation, that Frege and Tarski contradict each other with regard to the definability of truth then this is grossly misleading. For the expression 'definition of truth' is in this context *doubly equivocal*: the two opponents mean something different by the word 'truth' as well as by the word 'definition'. This is why Tarski's definition of truth is just as poorly suited to count against Frege's indefinability thesis as Frege's definition of I-truth would be. The relative truth that is defined in both of these cases must not be confused with the absolute truth whose indefinability is the subject of Frege's claim.

15.4.2   Frege's third and fourth arguments both rest on the omnipresence of the absolute truth of thoughts. Both arguments can also be directed against Tarski's definition of truth. The third argument can be used to show that any application of Tarski's concept of relative sentence truth is dependent on an application of the concept of absolute truth of thoughts. In accordance with the principle of omnipresence of (absolute) truth, a question 'p?' is synonymous with the question 'is it true that p?' Therefore, according to Frege the question

Is sentence p_0 true in G*?

would be synonymous with the following variant:

Is it (absolutely) true that p_0 is true in G*?

Here we see the dependence of sentence truth on the truth of thoughts, which is why sentence truth does not qualify as the kind of absolute and independent truth that Frege maintains is at the heart of science. This dependence, however, can be invoked in opposition to Tarski's definition of truth only if that definition is claimed to cover the proper, scientific concept of truth – for it is scientific truth that, according to Frege, cannot have a dependent nature. But why should the relative truth of sentences not be content with the status of a *secondary* kind of truth? Applications of a *secondary* kind of truth may legitimately depend on the primary, absolute kind of truth. If, on the other hand, Tarski's definition of truth is indeed meant as an explication of scientific truth, then the dependent character of relative sentence truth is an objection to it.

If Tarski's definition of truth in G* is intended as an explication of scientific truth, we can apply Frege's circle objection. Let us start with the following question:

Is sentence p_0 true?

Based on the aforementioned explication this question essentially amounts to:

Is sentence p_0 true in G*?

Now, when we apply the above definition of 'x is true in G*', we obtain the following questions:

(1) Is p_0 = 'Die Erde bewegt sich', and does the earth move?
(2) Is p_0 = 'Der Mond ist rund', and is the moon round?
(3) Is p_0 = [(S$_1$ und S$_2$)], and are both sentences true in G*?
(4) Is p_0 = [(Es ist nicht der Fall, dass S$_1$)] and is it not the case that S$_1$ is true in G*?

Due to the omnipresence of absolute truth these questions are synonymous with the following questions:

(1') Is p_0 = 'Die Erde bewegt sich', and *is it true that* the earth moves?

(2') Is p_0 = 'Der Mond ist rund', and *is it true that* the moon is round?

(3') Is p_0 = [(S$_1$ und S$_2$)], and *is it true that* both sentences are true in G*?

(4') Is p_0 = [(Es ist nicht der Fall, dass S$_1$)] and *is it true that* it is not the case that S$_1$ is true in G*?

(We could also add the expression 'is it true that' in the identity clauses, but I shall refrain from doing so for the sake of brevity.)

As (1') – (4') show, we then would need to already understand Frege's concept of truth in order to apply Tarski's definition of truth in G*. Thus, on the one hand we would have to already understand the explicandum (Frege's concept of truth) to apply the explicans (Tarski's concept of truth), but on the other hand we are supposed to first become acquainted with the precise meaning of the explicandum through the explicans. This is indeed a circle, and the circle shows that Tarski's definition of truth is not suited to explicating Frege's concept of truth.

In short: the attempt to understand Tarski's definition of truth as explicans of Frege's concept of truth fails due to (among other things) Frege's circle objection.

15.4.3 For Frege, the real conflict with Tarski concerns the logical status of the concept of truth: is truth absolute or relative? Frege's presumed relation to Tarski can be illustrated by means of the following sentence:

> It is (absolutely) true that the sentence 'Die Erde bewegt sich' is (relatively) true in G*.

There are two concepts of truth in this sentence: first Frege's ('It is (absolutely) true that p'), then Tarski's ('x is (relatively) true in G*'). The two concepts are related to each other in just the way Frege relates the scientific (absolute) concept of truth to the psychologistic (relative and hence 'improper') one: though we may use the *secondary*, relative concept of truth for certain purposes, in science – and especially in logic – only the primary, absolute concept of truth is authoritative.

As long as Tarski's relative truth refrains from attempting either to compete with Frege's absolute truth or to explicate it (and thereby to replace it in its role as proper truth), – in other words, as long as relative truth remains separate from and subordinate to absolute truth – both concepts can smoothly coexist. In this case we can follow Tarski in recursively defining the relative truth of sentences and at the same time follow Frege in denying that the absolute truth of thoughts could be made subject to an analytic definition. As long as the proper hierarchical ranking of the two concepts of truth remains in place there is, from Frege's point of view, objectively no conflict between the two theories of truth. This hierarchy can be briefly summarized as follows: Tarski's and Frege's concepts of truth do not stand in a relation of *explication* but in one of *subordination* one to the other.

15.4.4 Tarski does not generally rule out a coexistence of the two concepts of truth. He writes:

> [...] a time may come when we find ourselves confronted with several incompatible, but equally clear and precise, conceptions of truth. It will then become necessary to abandon the ambiguous usage of the word 'true,' and to introduce several terms instead, each to denote a different notion. Personally, I should not feel hurt if a future world congress of the 'theoreticians of truth' should decide – by a majority

of votes – to reserve the word *'true'* for one of the non-classical conceptions, and should suggest another word, say, *'frue,'* for the conception
considered here. But I cannot imagine that anybody could present
cogent arguments to the effect that the semantic conception is 'wrong'
and should be entirely abandoned. (Tarski, 1944, § 14, p. 28)

By 'incompatible, but equally clear and precise, conceptions of truth'
Tarski has in mind theories of truth that are able to present compara-
bly precise *definitions* of truth. But as Frege has shown with his circle
objection, this is not possible for absolute truth. Yet his requirements
for scientific truth – namely, that it must be absolute, perfect, and independent – are sufficiently 'clear and precise' to disqualify relative concepts of truth such as Tarski's or that of psychologism as inadequate. If
truth is absolute then Frege *does* 'present cogent arguments to the effect
that the semantic conception is "wrong" and should be entirely abandoned.' Now, if Tarski were content with defining the predicate 'x is frue
(!) in S' for sentences while acknowledging Frege's absolute truth as the
concept of truth that is prevalent in science, a peaceful coexistence of
their respective theories of truth would be possible. Though different in
various respects, the two theories need not contradict each other, as is
already obvious from the fact that the expression 'definition of truth' is
doubly equivocal in the context of the described conflict.

   The reason why Frege's theory of truth has scored so poorly in analytic
philosophy is that, over and over again, it has been treated mainly as a
foil to Tarski's. Tarski was regarded as Frege's great opponent, and since
Tarski's theory of truth prevailed, Frege's theory was to be rejected. But
if we neglect the mutually inconsistent *claims* that both were making to
explicate the proper or classic concept of truth, we realize that the two
theories are in many respects perfectly consistent with one another.
Hence, Frege's real opponent is not Tarski but only Tarski's *shadow*, that
is, a certain reception of Tarski's theory of truth that obscures Frege's
text to such an extent that one is no longer able to perceive his theory
of truth in any other way than as a caricature – specifically, a collection
of beginner's mistakes in logic. It is for this reason that in this book I
prefer to speak of a defense of Frege against Tarski's *shadow* rather than
against Tarski himself.

   A real conflict between Frege and Tarski arises only to the extent that
both theories aspire to adequately explicate *the same* concept of truth.
However, since Frege's object of explication is the absolute truth of
thoughts while Tarski's is the relative truth of sentences, the two theories are consistent in much the same way as Frege has shown the truth
of ideas and the truth of thoughts to be.

16
Conclusion I: Absolute versus Relative Truth

Frege's distinction between absolute and relative truth takes center stage in his argumentation in the third and fourth paragraphs of 'Der Gedanke'. In the following summary of the most important results I restrict myself to the consequences of this distinction – for my interpretation, for the common misinterpetations, and for the relation between Frege's and Tarski's respective theories of truth.

16.1 Absolute and relative truth

16.1.0 This book is concerned with Frege's theory of truth to the extent to which he develops it in the third and fourth paragraphs of 'Der Gedanke'. The key to understanding and defending Frege's argumentation is the distinction between *absolute* and *relative* truth.

If we make sure to distinguish strictly between the absolute truth of thoughts and the relative truth of ideas, we can provide a consistent reconstruction of Frege's argumentation. That is, indeed, the most important result of this book: Frege's argumentation *is* consistent.

A generalized version of Frege's argumentation can be briefly summarized as follows (see section 14.1):

(1) Scientific truth is absolute, that is, not relative, but correspondence truth is relative (relational).

(2) Scientific truth is perfect, but correspondence truth is imperfect.

(C1) Therefore, no relative (relational) truth with two different relata, and certainly not a correspondence truth, can qualify as scientific truth.

(C2) Therefore, the bearer of a relative truth (ideas or sentences) cannot qualify as bearer of scientific truth.

(3) Relative truth is dependent on the truth of thoughts and can be reduced to the latter. The truth of thoughts is independent and not reducible to any other truth.

(C3) The thought is the bearer of absolute truth. Scientific truth is the absolute truth of thoughts.

(4) The absolute truth of thoughts is indefinable because each attempt to apply an analytic definition of absolute truth fails due to a circle.

(5) If we relativize the truth of thoughts, for example, by interpreting such truth as correspondence then we enter into an infinite regress when trying to determine the truth bearer.

Again, Frege's conception of truth can be briefly summarized in one short sentence:

Truth is absolute, that is, not relative, and its bearer is the thought.

16.1.1 Four out of five arguments, namely arguments (1) – (3), and (5), each concern the *two* different concepts of truth. Frege here presupposes that scientific truth is absolute, perfect, and independent, and infers from this that the relative, imperfect, and dependent truth of correspondence cannot be scientific truth. Thus, Frege's critique of the correspondence theory is not an immanent form of critique, but an external one – one that rests on a *comparison* of the two concepts of truth.

The fourth argument concerns only *one* concept of truth. By means of a circle objection Frege here shows that absolute truth is *indefinable*. In the context of the other arguments, the circle objection is but a digression that can be smoothly integrated at that point because, just like the third argument, it presupposes the omnipresence of absolute truth.

16.1.2 Frege formulates three mutually complementary definitions of the correspondence truth of ideas, which together yield an adequate definition of the truth of ideas. This definition is *in itself* entirely unproblematic and not subject to a circle or regress objection. Frege rejects this definition only because it does not hold up in *comparison* with the requirements of scientific truth.

It is therefore misleading to mention Frege's critique of the correspondence theory in the *same* breath with his indefinability thesis. The external critique of the correspondence theory is directed against the relative truth of ideas and rests on a content-based comparison, while the indefinability thesis concerns only the absolute truth of thoughts and is justified by means of a formal, immanent circle objection.

16.1.3   The most important question in the third and fourth paragraphs of 'Der Gedanke' is the question of the *bearer of truth*. It is not by accident that the entire essay's title refers to the *thought* – the bearer of absolute truth.

The different truth concepts correspond to different bearers. The *idea* is the bearer of relative correspondence truth, while the *thought* is the bearer of absolute truth. The *sentence* occupies a middle position between idea and thought. Since the sentence expresses a thought depending on the respective context, it is related to the thought in one respect and to the idea in another, for the truth-values of sentences, just as those of ideas, are dependent on the context and hence relative. Accordingly, when reducing one truth concept to another Frege proceeds in two steps: he first reduces the truth of ideas to that of sentences, and subsequently sentence truth to the truth of thoughts.

16.1.4   Another important aspect of the difference between absolute and relative truth concerns the development of Frege's circle objection. In 'Der Gedanke' the circle objection is directed exclusively against the definability of the absolute truth of thoughts, while in the earlier parallel passage from 'Logik' it also targets the definition of the truth of ideas as correspondence. In 'Der Gedanke' Frege abandons the broad version of the circle objection he maintains in 'Logik'. The best explanation for this change in his argumentation is that in 'Der Gedanke' he differentiates more strictly between the truth of ideas and truth of thoughts than he does in 'Logik'. For the circle objection in the broad version becomes invalid once the strict distinction between the two concepts of truth comes into play.

16.2   The blind spot in the misinterpretations

Just as the distinction between absolute and relative truth is the key to understanding Frege's argumentation, so too the cause for the misinterpretations lies in the fact that this distinction has been overlooked in Frege's text and only *one* concept of truth is recognized therein. If Frege were discussing only one concept of truth, we would indeed be entitled to apply the relative concept of correspondence truth to the thought (the proper bearer of absolute truth) or, as Dummett does with his regress, to fuse Frege's critique of the correspondence theory of relative truth with his proof of the indefinability of absolute truth.

To Frege, the critique of the correspondence theory and the indefinability thesis are two distinct pairs of shoes, but in Tarski's shadow we are all too easily tempted to identify them with one another. For Tarski does claim that what he *defines* is the classic concept of truth, and that he does so precisely in the sense of the *correspondence theory*. But if both are fused in Tarski's account then does it surprise us that, under the impact of Tarski's theory of truth, one tends to conclude that Frege's indefinability thesis at the same time constitutes an attack against the correspondence theory?

It is also conceivable that the neglect of the distinction between the two concepts of truth in Frege's text is due to the fact that after Tarski absolute truth has been regarded as refuted, obsolete, and meaningless, so that such a concept should no longer be taken seriously in professional discussions. This would explain why Frege's external critique of the correspondence theory, which rests on a comparison between absolute and relative truth, has been conceived of as an immanent critique and why Frege has been credited with completely absurd arguments that barely have any foundation in his texts.

An important reason for the view that the absolute concept of truth has been refuted and is meaningless lies in Tarski's analysis of the liar antinomy. The absolute concept of truth can be indeed used to construct the liar antinomy, and therefore, Frege's concept of absolute truth seems to be threatened by the liar in a manner similar to that in which Frege's logic in the *Grundgesetze der Arithmetik* was threatened by Russell's antinomy. But this is not the case. Even if we want to follow Tarski in constructing a hierarchy of object languages and metalanguages to escape the liar, we are not thereby compelled to relativize the concept of truth. We could also construct a 'Fregean' hierarchy of languages and stick to the concept of absolute truth. Instead of saying 'Sentence x is (relatively) true on language level S_n' we could use the predicate 'The thought expressed by sentence x on language level S_n is (absolutely) true'. Thus, a Fregean hierarchy would suffice to escape the liar antinomy just as successfully as the Tarskian one.

In short: neither the liar antinomy nor Tarski's analysis of it has managed to prove Frege's concept of absolute truth to be obsolete.

16.3   Frege, Tarski and Tarski's shadow

16.3.0   Those who follow Tarski, and accept his definition of truth as an explication of the classic correspondence-based concept of truth, will tend to regard Frege's critique of the correspondence theory and his indefinability thesis as *fundamentally* incompatible with Tarski's theory of truth. Those who follow Tarski will probably reject Frege's argumentation.

Those, however, who follow Frege in distinguishing between absolute and relative truth will see that the two theories of truth do not necessarily exclude one another. Objectively speaking, there is only *one* genuine conflict between Frege and Tarski: if both claim that they explicate the same concept of truth by means of their respective theories, they are to that extent in conflict with one another. This conflict concerns the *claims* associated by them with their respective theories of truth. Nevertheless, what they actually *do* with their respective theories of truth is to a certain extent compatible. In the context of the conflict between Frege and Tarski the expression 'definition of truth' is in a twofold sense equivocal.

Frege specifies the relation between absolute and relative truth in two ways. On the one hand, he denies that relative truth could be truth in the proper sense. In this context Frege speaks of a *misuse* of the word 'truth', if thereby we mean the correspondence of ideas with something else. Frege himself always means by 'truth' absolute truth. In Tarski's aftermath, by contrast, the word 'truth' is always understood to mean relative truth, since absolute truth has been discarded as obsolete. Since in this book I deal both with Frege and with Frege scholars who have been influenced by Tarski, I here explicitly speak of *absolute* and *relative* truth.

On the other hand, Frege shows how the relative truth of ideas depends on the absolute truth of thoughts and can be reduced to the latter. The relation between Tarski's relative truth of sentences and Frege's absolute truth of thoughts can be determined analogously. But if we try to construe Tarski's definition of truth as an explication of Frege's absolute concept of truth, we will stumble over the circle objection.

Thus, Frege's distinction between relative and absolute truth also assists us in substantially alleviating the putative fundamental conflict between Frege's and Tarski's theories of truth. From this perspective, it was not Tarski himself who was responsible for obstructing the reception of Frege's theory of truth; it was, rather, Tarski's shadow.

16.3.1   As we can see, Frege's distinction between absolute and relative truth has profound significance for the investigations undertaken in this book in more than one respect.

Those who *overlook* the distinction tend to develop the view that Frege's critique is full of beginner's mistakes in logic and that Frege's and Tarski's respective theories of truth are contrary to one another.

But if we consistently *observe* the distinction then we will realize that Frege's argumentation in the third and fourth paragraphs of 'Der Gedanke' is consistent, and that Frege's and Tarski's respective theories of truth are not fundamentally incompatible.

17
Conclusion II: Two Modes of Interpretation

This book is intended to be an introduction to the practice of interpreting philosophical texts. By this I mean that a novice interpreter can take it as an example of the art of philosophical interpretation, using it to observe how an author engages in the practice of interpretation and in this way being introduced to that practice him- or herself. As with any art, it is more important to master the practice of interpreting than to be able to talk about it in the abstract and theoretically. Being able to say something smart and interesting about piano music might qualify you as a music critic, but it will hardly compensate for the miserable piano performance you gave the night before. This book is to introduce the reader to the practice, not the theory of interpretation (the latter of which is called 'hermeneutics').

That having been said, I do consider it worthwhile to briefly reflect at this point on the practice demonstrated in this book, now that we are at the end of our introduction to the art of interpreting. Every reader presumably understands by now that my interpretation differs from those of Dummett, Künne, Soames, and the rest – not just with regard to its length and results but especially with regard to the intensiveness with which I examined Frege's short texts. This illustrates the way in which interpretations can differ in *kind*, which raises the question: what kind of interpretation does this book introduce the reader to?

17.1 Six important rules that an interpretation should follow

17.1.0   Before going into this question I should like to discuss, as a result of this introduction to the art of philosophical interpretation, six important requirements that characterize my interpretation methodologically.

Let me start with four requirements that should be taken into account in every good interpretation, and illustrate by way of examples some questions that have arisen with respect to these requirements (cf. Chapter 13). These examples should also make it clear once more just how important it is for an interpretation to follow the aforementioned rules and how closely they are interconnected.

(I) *Demarcation*: Demarcation of the entire string of arguments from its context as well as demarcation of the individual arguments within the string from one another.

How should we demarcate Frege's critique of the correspondence theory? How should we demarcate his argument for the indefinability of truth? Does the fourth paragraph still continue the argumentation presented in the third paragraph? Is the third argument linked more closely to the second or rather to the fourth? Are the third and fourth arguments essentially one and the same, or are they two distinct arguments? Is the parallel passage in 'Logik', in which the first and second arguments are missing and basically only the third and fourth arguments – combined into *one* argumentative unit – occur, evidence that the two arguments belong together also in 'Der Gedanke'? Or is, conversely, the addition of the first and second arguments in 'Der Gedanke' evidence that Frege's way of arguing has changed and that therefore the allocation of the arguments to one another has changed as well?

(II) *Emphasis*: Assignment of emphasis to the various parts of the argumentation as well as to the central claims and concepts.

Is the *indefinability thesis* in the third paragraph of 'Der Gedanke' the central thesis of the first four paragraphs, or is it rather Frege's specification of the *thought* as the primary bearer of truth in the fourth paragraph? How do the two claims relate to one another: Is the fourth argument merely a detour in the context of specifying the primary bearer of truth, or are the first two arguments merely an irrelevant prefix to the proof of indefinability? Are the concepts of a picture, of an idea, of a sentence and a thought as possible bearers of truth irrelevant for Frege's arguments? Are they interchangeable at random without affecting the soundness of Frege's arguments?

(III) *Concepts*: The interpretation of the concepts, especially with respect to possible deviations from our contemporary terminology.

Does Frege mean by a relation word in 'Der Gedanke' the same as what we mean by a two-place predicate? Or is a relation word, as this term is used by Frege in this context, rather a one-place predicate designating a relational property? Does Frege use only one concept of truth in his argumentation that we could apply without any modifications to various types of truth bearers, or does he distinguish between different concepts of truth? Does Frege employ the notion of correspondence truth with respect to sentences or thoughts, as we do today, or does he employ this notion only in application to images and ideas? Are the notions of a circle and a regress interchangeable for Frege? Is a circle in definition the same for Frege as a circle in the application of a definition?

(IV) *Issues*: Identification of the questions that the text is intended to answer, and the assignment of a central topical issue to the text.

Is Frege primarily concerned in his first four arguments with the question of the *indefinability* of truth, or rather with the questions of the proper *bearer* of truth? The answer to this question determines how we read, demarcate and interpret Frege's text. Conversely, we arrive at an answer to this question only by working through the other requirements. All four rules are most closely connected, for by working on one of them we automatically also address the others.

17.1.1　Now, if we discuss or refer to other interpretations in the context of our own then we need to follow two additional rules.

(V) *Quotations*: Reviewing of the quotations that occur in the other interpretations (if any) with respect to the demarcation of the respective quotes as well as the omissions within the quotes.

Künne's Frege quotes clearly show that he demarcates the third and fourth arguments as one unit from the rest of the text and that he replaces the expressions 'picture' and 'idea' by dots, thereby indicating that he regards them as irrelevant for the arguments. Such quotes clearly reveal the interpreter's demarcation of and assignment of emphasis within the text. We can then ask whether it might not be possible to read the text differently.

(VI) *Contraposition*: The contraposition of an absurd interpretation, which we can thereby use to develop an alternative interpretation.

In Chapter 13, I used two examples to show how this rule can be applied in practice and how productive its application can be. Since from a didactic point of view this rule is an important result of this book, I should like to repeat it here once more explicitly:

> If demarcating the arguments, assigning emphasis to different parts of the text, interpreting the concepts, and assigning the text an overall topical issue – as this has been done in the particular interpretation under our scrutiny – together result in a completely absurd argument, then this absurd argument can be explained as a misunderstanding or misinterpretation only if we succeed in demarcating the arguments in a different way, putting different emphases on the various parts of the text, reinterpreting some concepts differently, or assigning the text a different topical issue than was done in that interpretation.

17.1.2 To complete this section I would like to briefly demonstrate now how closely rules 1–4 are interconnected and how much they affect one another. It almost does not matter which rule we apply first. However, the connection may reveal itself most clearly if we start by assigning the text its overall topical issue.

If we read Frege's text primarily under the question of the definability of truth, then we need not consider the fourth paragraph, because it does not contribute anything to answering this question. In this case we will not understand the fourth argument as a detour in a very different argumentative context, but as the most important goal of the argumentation in the third paragraph. Without the fourth paragraph, however, Frege's distinction between the different concepts of truth becomes weakened, so that we can more easily overlook the fact that Frege distinguishes at least two concepts of truth and that for him the question of the bearer of scientific truth is more important in this context. If we then go on to regard the third and fourth arguments in 'Der Gedanke' as constituting the decisive answer to the question of definability, just as in the parallel passage in 'Logik', then we will end up taking the two arguments to be essentially one and the same. This identification of the two arguments, however, will block off any appropriate interpretation of either of them and almost inevitably lead to one or other variation of Dummett's regress (cf. section 13.3).

Even though all four rules are, strictly speaking, equally important for an interpretation, I do consider the *demarcation* of the text and the assignment of a central *topical issue* as the most crucial. If we manage to accurately demarcate the text and grasp the overall topical issue with which it deals, then we should be able to accurately interpret the arguments in the text – at least if the text is internally consistent, as it is the case with Frege's text.

17.1.3   With that I should like to conclude this introduction to the interpretation of philosophical texts. In this book I have shown how we can interpret a text as consistent piece of argumentation on condition that the text actually is internally consistent. I have not discussed the different question of how to interpret inconsistent texts, which certainly exist, and I did not actually have any opportunity to do so, since Frege's text turned out to be consistent. For this reason I shall not go any further into discussing the interpretation of inconsistent texts, since this would be a topic beyond the remit of this book.

Whether a text is consistent or not can be decided only after one or more attempts at an interpretation; and even then the result will be tentative. If an interpretation does not match the text, then this may mean one of two things. We have either interpreted away the contradictions inherent in an internally inconsistent text or, if the text is internally consistent, have interpreted inconsistencies into it that it does not really have. It goes without saying that we should avoid both of these mistakes. Since, however, present-day analytic philosophy is more prone to interpreting inconsistencies into consistent texts I should like to spend some more time on this matter.

If my interpretation is correct then the other interpreters made the mistake of rushing to condemn Frege's argumentation as inconsistent. The following considerations are meant to critically address such rushed condemnations of philosophical texts as logically inconsistent. In particular, such condemnations are rushed if they are not preceded by a thorough investigation of the text itself. But until now a thorough investigation of Frege's critique of the correspondence theory has not taken place, and I regard the manner in which scholars have made fun of Frege's alleged beginner's mistakes in logic as scandalous. That is why the following considerations are formulated not just with the subject matter in mind but also spiced with a considerable pinch of polemic.

In what follows I shall not offer a complete typology of interpretations (that is, it is *not* my aim to engage in a discussion of hermeneutics); rather, I shall merely distinguish two important, and radically different, kinds or tendencies of interpretation, namely *apologetic* and *incriminating* interpretations. This book is intended as an example of apologetic interpretation, while the alternative interpretations discussed are closer to incriminating interpretations.

17.2 Apologetic interpretations

17.2.0 Originally, 'apologetics' meant the justification of Christian doctrines. However, I use the expression in a different, weaker sense. Apologetic interpretation as I use the term is not concerned with justifying theses proposed in the text being interpreted, nor, obviously, is it confined to any specific doctrine. Rather, the goal of interpreting a philosophical text apologetically should be solely the reconstruction of an argument, or a series of arguments, as *internally consistent* (self-consistent). By showing the text to be internally consistent, we demonstrate that it satisfies a crucial requirement for qualifying as a worthwhile topic of contemporary philosophical discussion as well. The interpreter should use the concepts as the author intended them to be used, and he should use the same presuppositions and present the author's arguments as clearly and convincingly as possible. He should presuppose that the arguments in the text are consistent and subsequently attempt to find a reconstruction in which the text is indeed consistent, after which he should defend the consistency of the arguments against overly rash attacks. This is what I mean when I say that the interpretation offered in Chapters 1–12 is 'apologetic'.

Whether the premises or conclusions asserted in these arguments are true is a question that reaches beyond the framework of a mere interpretation. When we interpret a text apologetically, we can either agree or disagree with the premises or conclusions of the interpreted text, or we can abstain from any judgment as to their truth or falsity. Hence an apologetic interpretation should be distinguished from a judgment or reaction concerning the content of the text – much as we need to distinguish between the lawyers and the judge in a court setting. The lawyer's task, for example, is to represent his client. If the client's concern or pleading is dismissed in court on formal grounds, the lawyer has to share the blame for this. Likewise, if a philosophical text is excluded from professional discussion on the grounds that it is obviously inconsistent then the interpreter often should share the blame. Just as an absent client needs his lawyer to represent him in court, a deceased philosopher needs good interpreters who can direct the attention of contemporary philosophical discussions to his or her arguments. The impression that a text should be rejected on formal grounds (that is, as 'self-contradictory' or 'logically invalid') is just as fatal in the context of philosophical discussion as a formal defect in court, for the result will be that no one regards that text as worth spending any time on.

Even a text that has been interpreted as being internally consistent may be rejected without much discussion if it appears to contradict some basic principles of contemporary philosophy, for this would make it *externally* inconsistent. By showing that this contradiction is only a seeming one, and hence that the text is externally consistent and so still constitutes a serious contribution to discussions of the topic, we defend the text in a more comprehensive manner than we would if we had focused only on proving that it is internally consistent. This is how I defended Frege's arguments in Chapters 14–15 against the common objection that the liar antinomy and Tarski's conception of truth force us to abandon Frege's absolute concept of truth. I call such a defense (or 'apology'), which establishes both internal and external consistency, an 'apologetic interpretation in the broad sense'. In the following I mean by an apologetic interpretation mostly this broader sense of the term.

It is an apologetic interpreter's task to draw attention to a text's statement and to introduce the text as a relevant contribution to contemporary discussion. This is why I have tried to show that the concept of absolute truth is not meaningless or absurd. Even if I have succeeded in showing this, I have not yet shown that the absolute concept of truth is the 'right' concept of truth. That appears to be Frege's thesis, and I consider it worth discussing. Yet I have not attempted to positively argue that this thesis is true; instead, I merely defended it as not being straightforwardly absurd. In other words, I have not tried to establish that Frege's thesis is true, but merely to refute a potential objection to it. In short: I have offered an apologetic interpretation in the broad sense.

An even more comprehensive 'apology' would consist in defending the premises and conclusions of the text as true. In my view, however, this would no longer constitute a mere defense but amount to a justification corresponding to theological apologetics. In my view, such a justification is beyond the apologetic interpretation of a philosophical text.

17.2.1 Of course, an attempted apologetic interpretation can fail; at that point it has to be abandoned. In this case the interpreter, despite her or his intensive efforts, has not succeeded in interpreting the text as self-consistent; she or he must, then, give up her or his initial presumption that the text is consistent. When all attempts fail, it sometimes remains open to dispute whether this was the interpreter's shortcoming or indeed due to ineliminable inconsistency in the text itself. As interpreters we should always keep the possibility in mind that our attempts at an accurate and consistent interpretation may fail, even though the text 'itself' may not be inconsistent and may allow for a consistent reconstruction. Thus, we must be careful to not blame the text for our own possible limitations. Though in this way we will be circumspect with our assessment of a text as inconsistent, we undoubtedly acknowledge that there are numerous inconsistent texts (even though it may be subject to dispute precisely which texts these are).

I do recognize the danger here that we might be tempted to succumb to the other extreme of smoothing inconsistent texts over – that is, interpreting them as consistent. However, I shall not address this problem here any further, because contemporary analytic philosophy strikes me as being much closer to the opposite extreme of interpreting consistent texts as inconsistent if they promote views with which the interpreters do not agree.

An apologetic interpretation requires that as long as we are in the process of interpreting a text and have not found a reason to stop that process, we adhere to the assumption that the text is self-consistent. It is all a matter of finding a point of view from which the text can be reconstructed as a consistent and comprehensible argumentation. The process of apologetic interpretation consists of the following main steps, which are closely linked to one another (see above): reading the text, demarcating it, assessing it, paraphrasing its concepts, considering it as a response to a certain question, and all this over and over again, patiently and from various angles until the text has been understood – or until the attempt is given up. In short, apologetic interpretation consists in a slow, thorough and repeated reading of a text, and in thinking about it 'day and night', so to speak. Or, put more simply: it consists in *reading* the text.

17.3 Incriminating interpretations

17.3.0 One often finds in philosophical disputes that rival texts are mostly interpreted not in a spirit of a defense but in one of prosecution; they are interpreted, that is, not apologetically but *incriminatingly*. Incriminating interpretations can prevent a reader from reading the original text thoroughly because they convey the impression that reading the text is not worthwhile in the first place. Usually, the incriminating interpretation serves to represent the text as internally inconsistent and absurd. The reader is thus encouraged to do without engaging with the original text altogether, since the opponent has been revealed to contradict him- or herself, which renders his or her arguments unsound on logical grounds and, as a self-contradictory position, ultimately incomprehensible. Alternatively, sometimes the text is presented as being externally inconsistent – that is, as openly contradicting some generally acknowledged truths – and hence as unacceptable. Incriminating interpretations tend to distort the interpreted text until it becomes a caricature of itself.

Within the tradition of analytic philosophy the paradigm of an incriminating interpretation is Carnap's notorious critique of Heidegger. One of Carnap's main objections is that Heidegger conjures up his metaphysical 'Nothingness' by some sleight-of-hand, namely by inadmissible nominalization of the logical 'nothing' ('¬∃x'). Carnap's incriminating interpretation has contributed to the fact that even today, few if any analytic philosophers have ever read Heidegger's text – or even just the passages quoted by Carnap – carefully and thoughtfully. If they had, then they could not have missed the fact that for Heidegger the metaphysical nothingness is not preceded by the logical 'nothing', as Carnap had maintained, but the expression 'nothing else', which by no means can be interpreted as a negated existential quantifier in Heidegger's argumentative context. This small difference between 'nothing' and 'nothing else' undermines Carnap's critique. This could have been easily detected by anyone who had carefully read the Heidegger passage quoted by Carnap and compared it to Carnap's interpretation. Alas, such a re-examination did not take place until 70 years later (see Pardey, 2006, chapter IV).

A crucial factor in the non-occurrence of any such re-examination was, in my view, Carnap's incriminating interpretation, which prevented a careful reading of the original text. Carnap conveys the impression that it is not worthwhile reading Heidegger's text, since whoever nominalizes the logical 'nothing', that is the negated existential quantifier, and thus attempts to use it as basis for a metaphysics of 'nothingness' must have written sheer nonsense! In short, the effect of an incriminating interpretation is as follows: the original text is *no* longer read, or at least not read carefully, and readers are content with studying the interpretation instead.

17.3.1 The following are the main characteristics of an incriminating textual interpretation:

(1) There is asserted to be a direct contradiction between the text and a contemporary dogma. The mere fact that the text contradicts the

dogma alone already raises the suspicion that the arguments in the text are absurd.

(2) The very argument in the text that runs counter to the dogma is said to be completely absurd.

(3) A quoted passage is presented and is claimed to demonstrate that the text contains this absurd argument.

(4) Careful reading of the original text reveals that the assessment established in (1) – (3) rests on a misinterpretation.

Carnap's representation of Heidegger displays all four of these characteristics:

(1) The contradiction with received dogma: According to logical positivism all metaphysics is meaningless, and in particular a metaphysics of nothingness. Yet Heidegger's text presents this very meaningless, metaphysical nothingness as a meaningful philosophical concept.

(2) The absurd argument: The logical 'nothing' is nothing but the negated existential quantifier '¬∃x'. The nominalization of 'nothing' does not establish an object that we could baptize 'the Nothingness', but only gibberish.

(3) The quoted passage: Heidegger performs this nominalization by 'Only the Being is supposed to be examined, but nothing else. [...] What about this Nothingness?' (Heidegger, 1929, p. 28)

(4) The misinterpretation: Contrary to Carnap's assessment, Heidegger does not make a transition from the logical 'nothing' to the metaphysical 'nothingness', but from the expression 'nothing else', which cannot be interpreted as a negated existential quantifier in Heidegger's context.

17.4 Refutation by way of interpretation

17.4.0 The controversy between Carnap and Heidegger represented two contrary traditions in philosophy that clashed violently. Another historical example of a battle between two traditions is Ryle's critique of Descartes. Ryle's interpretation of Descartes must in part come across as positively alien to anyone who has taken the time to read Descartes carefully instead of relying on Ryle's representation of him. Some modern logic criticisms of classical Aristotelian logic also rest on an incriminating interpretation. And there are many other controversies in philosophy in which criticisms of the opponent rest to a large extent on blunt misinterpretations (see Pardey, 2006, chapters II, IV–VI).

Though blunt errors of interpretation are not acceptable – after all, we would expect philosophers to be able to read! – we do tend to forgive them in the context of intense controversies between different schools of philosophy, and especially if they occur in polemical pamphlets such as Carnap's 'The Elimination of Metaphysics through Logical Analysis of Language'. Such writings focus less on logic, better arguments and the goal of convincing the reader rationally than on the use of rhetoric and the art of persuasion. Carnap's critique of Heidegger was faulty in terms of logical analysis, yet it is a superb example of the *rhetorical* use – or abuse! – of logical concepts.

The controversy between Frege and Tarski concerning the concept of truth, however, is not an example of a clash of different schools but a dispute within the school of analytic philosophy itself. It is a shocking testimony on the state of analytic philosophy that Frege's critique of the correspondence theory has been received for 40 years as a mere collection of beginner's mistakes in logic and that it was possible to present Frege as the clown among scholars of the theory of truth, even though this assessment rested on blunt misinterpretations. Has analytic philosophy become so dogmatic that arguments deviating in crucial respects from contemporary established dogmas are recognized only as caricatures?

17.4.1 Though Dummett, Künne, Soames and so on in general have a positive attitude toward Frege, their interpretations – critically examined in this book – come dangerously close to incriminating interpretation. I shall briefly demonstrate this by reference to their accounts of Frege's first and his third/fourth arguments.

Regarding the first argument, the presence of the four characteristics of an incriminating interpretation can be illustrated as follows:

(1) The contradiction with received dogma: As Tarski has shown, the concept of truth can be defined in terms of the correspondence theory of truth. According to Frege truth is not definable in terms of the correspondence theory of truth.
(2) The absurd argument: One-place concepts are not two-place concepts, which is why the one-place predicate 'x is true' cannot be defined by means of the two-place predicate 'x corresponds to y'.
(3) The quoted passage: 'It might be supposed from this that truth consists in a correspondence [...]. Now a correspondence is a relation.

But this goes against the use of the word "true", which is not a relation word.'

(4) The misinterpretation: Contrary to what has been claimed in interpretations still dominant today, Frege denies that truth is a relational property.

With regard to the third/fourth arguments the four characteristics are
exemplified as follows:

(1) The contradiction with received dogma: The concept of truth
can be defined (according to Tarski). According to Frege truth is
indefinable.
(2) The absurd argument: See Dummett's regress (Chapter 10).
(3) The quoted passage: 'And with that we should be confronted again
by a question of the same kind, and the game could start all over.'
(4) The misinterpretation: Neither the third nor the fourth argument is
a regress argument. By the third argument Frege shows the dependency of I-truth on S-truth, and the fourth argument is a circle
argument.

It is one thing to interpret an argument, and quite another to refute it.
Interpreting and refuting are two distinct activities, just as the interpretation of a text is distinct from the justification of its theses. An
incriminating interpretation, however, conflates the two activities,
thereby abbreviating the work process. This is as if a plaintiff in court
is at the same time the judge, who can thus sentence the defendant
at the moment at which the charges are filed. Thus, for example, in
Künne's and Soames's interpretations Frege's first argument is presented
in such a manner that its alleged absurdity becomes immediately obvious to any reader. The interpreters thus use the interpretation itself to
pronounce a verdict on the argument. Since the argument basically is
refuted already by way of its interpretation, we could also simply call
the incriminating interpretation the method of 'refutation by way of
interpretation'.

17.5   On the uses and the morals of interpreting

Before concluding this book I should like to briefly comment on the
uses and the morals of interpreting. Why should we learn how to adequately interpret texts in the first place? Could we not do very well in
life without this type of art?

It is a truism that human life is not conceivable without communication between people, and communication does not exist without interpretation of the utterances of others. Unfortunately, we do not always have the time or leisure to adequately interpret the utterances of others; as a result, we often make situations worse due to our misinterpretations. It is therefore even more important that at least in theoretical contexts, such as in philosophy classes at college or graduate school, the art of interpreting be accurately studied. For only then could this art perhaps, after some training and practice, also be applied in some everyday situations. To put it bluntly: we do not learn interpreting just for college or graduate school but also for life!

It therefore matters greatly whether we engage in an apologetic or an incriminating interpretation of an opponent's text. Incriminating interpretations constitute a 'refutation by way of interpretation' and are usually unfair to the original text; they lack respect for their opponent and signal dogmatism and intolerance, favoring a 'tunnel vision'. Tolerance and fairness require that we allow that our opponent's arguments may be consistent and comprehensible, even if we reject them for reasons whether they be objective or personal. Fair interpretations are always apologetic. (Please note that I do *not* use the term 'apologetics' in the sense of 'justification'.)

Western culture prides itself on being a culture of freedom. The freedom of thought consists in being able to think differently from the majority of people, from the authorities, and from the spirit of the time. If now even analytic philosophy reinterprets deviant thinking as absurd thinking – as has been done with Frege's critique of the correspondence theory – then I wonder about our Western culture's past and present attitude toward other cultures in general.

If now the reader, realizing his or her own powerlessness, is inclined to resign in the light of the so-called 'clash of cultures', I would like to add that we can avoid such resignation by applying some kind of contraposition as follows. If we wish to improve our relations with other cultures then we should learn to interpret the texts and utterances of others apologetically even if they deviate from our beliefs. It may not be much, but it is a step in the right direction!

Appendix: Synopsis of Relevant Frege Passages and Their Various Translations

The translations of the German original texts largely follow the translations by Geach/Stoothoff and Long/White, respectively. However, in some places we (Lotter/Pardey) have found it necessary to slightly alter those translations. The most important alterations (1) – (9) are commented in section A.3.

A.1 'Der Gedanke'

Table A.1 The third and fourth paragraphs of 'Der Gedanke'

Frege: 'Der Gedanke'	Translation by Geach and Stoothoff	Modified translation
Das Wort ‚wahr' erscheint sprachlich als Eigenschaftswort. Dabei entsteht der Wunsch, das Gebiet enger abzugrenzen, auf dem die Wahrheit ausgesagt werden, [(1)] wo überhaupt Wahrheit in Frage kommen könne.	Grammatically, the word 'true' looks like a word for a property. So we want to delimit more closely the region within which truth can be predicated, the region [(1)] in which there is any question of truth.	Grammatically, the word 'true' appears as an adjective. Hence, the desire arises to delimit more closely the region within which truth could be predicated, the region [(1)] in which the question 'Is it true?' could be in principle applicable.
Man findet die Wahrheit ausgesagt von Bildern, Vorstellungen, Sätzen und Gedanken. Es fällt auf, daß hier sichtbare und hörbare Dinge zusammen mit Sachen vorkommen, die nicht mit den Sinnen wahrgenommen werden können.	We find truth predicated of pictures, ideas, sentences, and thoughts. It is striking that visible and audible things turn up here along with things which cannot be perceived with the senses.	We find truth predicated of pictures, ideas, sentences, and thoughts. It is striking that visible and audible things occur here along with things which cannot be perceived with the senses.
Das deutet darauf hin, daß Verschiebungen des Sinnes vorgekommen sind.	This suggests that shifts of meaning have taken place.	This suggests that alterations in sense have taken place.

Continued

Table A.1   Continued

Frege: 'Der Gedanke'	Translation by Geach and Stoothoff	Modified translation
In der Tat! Ist denn ein Bild als bloßes sichtbares, tastbares Ding eigentlich wahr? und ein Stein, ein Blatt ist nicht wahr?	So indeed they have! Is a picture considered as a mere visible and tangible thing really true, and a stone or a leaf not true?	Indeed they have! For is a picture, as a mere visible and tangible thing, really true? And a stone, a leaf is not true?
Offenbar würde man das Bild nicht wahr nennen, wenn nicht eine Absicht dabei wäre. Das Bild soll etwas darstellen.	Obviously we could not call a picture true unless there were an intention involved. A picture is meant to represent something.	Obviously, we would not call a picture true unless there were an intention involved. A picture is meant to represent something.
Auch die Vorstellung wird nicht an sich wahr genannt, sondern nur im Hinblick auf eine Absicht, daß sie mit etwas übereinstimmen solle.	(Even an idea is not called true in itself, but only with respect to an intention that the idea should correspond to something.)	Neither is an idea called true in itself, but only with respect to an intention that it should correspond to something.
Danach kann man vermuten, daß die Wahrheit in einer Übereinstimmung eines Bildes mit dem Abgebildeten bestehe.	It might be supposed from this that truth consists in a correspondence of a picture to what it depicts.	It might be supposed from this that truth consists in a correspondence of a picture to what it depicts.
Eine Übereinstimmung ist eine Beziehung.	Now a correspondence is a relation.	Now a correspondence is a relation.
Dem widerspricht aber die Gebrauchsweise des Wortes ‚wahr', das kein Beziehungswort ist, keinen Hinweis auf etwas anderes enthält, mit dem etwas übereinstimmen solle.	But this goes against the use of the word 'true', which is not a relative term and contains no indication of anything else to which something is to correspond.	But this goes against the use of the word 'true', which is not a relation word, does not contain any indication of anything else to which something is to correspond.
Wenn ich nicht weiß, daß ein Bild den Kölner Dom darstellen solle, weiß ich nicht, womit ich das Bild vergleichen müsse, um über seine Wahrheit zu entscheiden.	If I do not know that a picture is meant to represent Cologne Cathedral then I do not know what to compare the picture with in order to decide on its truth.	If I do not know that a picture is meant to represent Cologne Cathedral then I do not know what to compare the picture with in order to decide on its truth.
Auch kann eine Übereinstimmung ja nur dann vollkommen sein, wenn die übereinstimmenden Dinge zusammenfallen, also gar nicht verschiedene Dinge sind.	A correspondence, moreover, can only be perfect if the corresponding things coincide and so just are not different things.	A correspondence, moreover, can only be perfect if the corresponding things coincide and so just are not different things.

Continued

Table A.1 Continued

Frege: 'Der Gedanke'	Translation by Geach and Stoothoff	Modified translation
Man soll die Echtheit einer Banknote prüfen können, indem man sie mit einer echten stereoskopisch zur Deckung zu bringen sucht.	It is supposed to be possible to test the genuineness of a bank-note by comparing it stereoscopically with a genuine one.	It is supposed to be possible to test the genuineness of a banknote by comparing it stereoscopically with a genuine one.
Aber der Versuch, ein Goldstück mit einem Zwanzigmarkschein stereoskopisch zur Deckung zu bringen, wäre lächerlich. Eine Vorstellung mit einem Dinge zur Deckung zu bringen, wäre nur möglich, wenn auch das Ding eine Vorstellung wäre.	But it would be ridiculous to try to compare a gold piece stereoscopically with a twenty-mark note. It would only be possible to compare an idea with a thing if the thing were an idea too.	But it would be ridiculous to try to compare a gold piece stereoscopically with a twenty-mark note. It would only be possible to compare an idea with a thing if the thing were an idea too.
Und wenn dann die erste mit der zweiten vollkommen übereinstimmt, fallen sie zusammen.	And then, if the first did correspond perfectly with the second, they would coincide.	And then, if the first corresponds perfectly with the second, they coincide.
Aber das will man gerade nicht, wenn man die Wahrheit als Übereinstimmung einer Vorstellung mit etwas Wirklichem bestimmt.	But this is not at all what people intend when they define truth as the correspondence of an idea with something real.	But this is not at all what people intend when they define truth as the correspondence of an idea with something real.
Dabei ist es gerade wesentlich, daß das Wirkliche von der Vorstellung verschieden sei.	For in this case it is essential precisely that the reality shall be distinct from the idea.	For in this case it is essential precisely that the real thing be distinct from the idea.
Dann aber gibt es keine vollkommene Übereinstimmung, keine vollkommene Wahrheit. Dann wäre überhaupt nichts wahr; denn was nur halb wahr ist, ist unwahr.	But then there can be no complete correspondence, no complete truth. So nothing at all would be true; for what is only half true is untrue.	But then there can be no perfect correspondence, no perfect truth. So nothing at all would be true; for what is only half true is untrue.
Die Wahrheit verträgt kein Mehr oder Minder.	Truth does not admit of more or less. – [!]	Truth does not admit of more or less.
[(2)] Oder doch? Kann man nicht festsetzen, daß Wahrheit bestehe, wenn die Übereinstimmung in einer gewissen Hinsicht stattfinde?	[(2)] But could we not maintain that there is truth when there is correspondence in a certain respect?	[(2)] Or does it? Could we not maintain that there is truth when there is correspondence in a certain respect?
Aber in welcher [Hinsicht]?	But which respect?	But which respect?
Was müßten wir dann aber tun, um zu entscheiden, ob etwas wahr wäre?	For in that case what ought we to do so as to decide whether something is true?	And what would we then have to do so as to decide whether something were true?

Continued

Table A.1 Continued

Frege: 'Der Gedanke'	Translation by Geach and Stoothoff	Modified translation
Wir müßten untersuchen, ob es wahr wäre, daß – etwa eine Vorstellung und ein Wirkliches – in der festgesetzten Hinsicht übereinstimmten.	We should have to inquire whether it is *true* that an idea and a reality, say, correspond in the specified respect.	We should have to inquire whether it were true that, for example, an idea and something real correspond in the specified respect.
[(3)] Und damit ständen wir wieder vor einer Frage derselben Art, und das Spiel könnte von neuem beginnen.	[(3)] And then we should be confronted by a question of the same kind, and the game could begin again.	[(3)] And with that we should be confronted again by a question of the same kind, and the game could start all over.
So [(4)] scheitert dieser Versuch, die Wahrheit als eine Übereinstimmung zu erklären.	So [(4)] the attempted explanation of truth as correspondence breaks down.	So [(4)] this attempt to explain truth as correspondence breaks down.
[(5)] So scheitert aber auch jeder andere Versuch, das Wahrsein zu definieren.	[(5)] And any other attempt to define truth also breaks down.	[(5)] But likewise, any other attempt to define truth also breaks down.
Denn in einer Definition gäbe man gewisse Merkmale an.	For in a definition certain characteristics would have to be specified.	For in a definition certain characteristics would have to be specified.
Und bei der Anwendung auf einen besonderen Fall käme es dann immer darauf an, ob es wahr wäre, daß diese Merkmale zuträfen. So drehte man sich im Kreise.	And in application to any particular case the question would always arise whether it were *true* that the characteristics were present. So we should be going round in a circle.	And in application to any particular case it would always depend on whether it were true that the characteristics were present. So we should be going round in a circle.
Hiernach ist es wahrscheinlich, daß der Inhalt des Wortes ‚wahr' ganz einzigartig und undefinierbar ist.	So it seems likely that the content of the word 'true' is *sui generis* and indefinable.	So it is likely that the content of the word 'true' is sui generis and indefinable.
Wenn man Wahrheit von einem Bilde aussagt, will man eigentlich keine Eigenschaft aussagen, welche diesem Bilde ganz losgelöst von anderen Dingen zukäme, sondern man hat dabei immer noch eine ganz andere Sache im Auge und man will sagen, daß jenes Bild mit dieser Sache irgendwie übereinstimme.	When we ascribe truth to a picture we do not really mean to ascribe a property which would belong to this picture quite independently of other things; we always have in mind some totally different object and we want to say that the picture corresponds in some way to this object.	When we predicate truth of a picture we do not really mean to predicate a property which would belong to this picture altogether independently of other things. Rather, we always have in mind some totally different object and we want to say that that picture corresponds in some way to this object.

Continued

Table A.1   Continued

Frege: 'Der Gedanke'	Translation by Geach and Stoothoff	Modified translation
‚Meine Vorstellung stimmt mit dem Kölner Dome überein' ist ein Satz, und es handelt sich nun um die Wahrheit dieses Satzes.	'My idea corresponds to Cologne Cathedral' is a sentence, and now it is a matter of the truth of this sentence.	'My idea corresponds to Cologne Cathedral' is a sentence, and now it is a matter of the truth of this sentence.
So wird, was man wohl mißbräuchlich Wahrheit von Bildern und Vorstellungen nennt, auf die Wahrheit von Sätzen zurückgeführt.	So what is improperly called the truth of pictures and ideas is reduced to the truth of sentences.	So what is improperly called the truth of pictures and ideas is reduced to the truth of sentences.
Was nennt man einen Satz? Eine Folge von Lauten; aber nur dann, wenn sie einen Sinn hat, womit nicht gesagt sein soll, daß jede sinnvolle Folge von Lauten ein Satz sei.	What is it that we call a sentence? A series of sounds, but only if it has a sense (this is not meant to convey that *any* series of sounds that has a sense is a sentence).	What is it that we call a sentence? A series of sounds, but only if it has a sense, which is not to say that any series of sounds that has a sense is a sentence.
Und wenn wir einen Satz wahr nennen, meinen wir eigentlich seinen Sinn.	And when we call a sentence true we really mean that its sense is true.	And when we call a sentence true we really mean that its sense is true.
Danach ergibt sich als dasjenige, [(1)] bei dem das Wahrsein überhaupt in Frage kommenkann, der Sinn eines Satzes.	And hence the only thing [(1)] that raises the question of truth at all is the sense of sentences.	And hence the only thing [(1)] for which the question 'Is it true?' can be in principle applicable is the sense of sentences.
Ist nun der Sinn eines Satzes eine Vorstellung?	Now is the sense of the sentence an idea?	Now is the sense of the sentence an idea?
Jedenfalls besteht das Wahrsein nicht in der Übereinstimmung dieses Sinnes mit etwas anderem; denn sonst wiederholte sich [(6)] die Frage nach dem Wahrsein ins Unendliche.	In any case, truth does not consist in correspondence of the sense with something else, for otherwise [(6)] the question of truth would get reiterated to infinity.	In any case, truth does not consist in correspondence of the sense with something else, for otherwise [(6)] the question of whether something is true would get reiterated to infinity.

A.2 'Logik'

Table A.2 The parallel passage in 'Logik'

Frege: 'Logik'	Translation by Long and White	Modified translation
Es wäre nun vergeblich, durch eine Definition deutlicher zu machen, was unter ,wahr' zu verstehen sei.	Now it would be futile to employ a definition in order to make it clearer what is to be understood by 'true'.	Now it would be futile to employ a definition in order to make it clearer what is to be understood by 'true'.
Wollte man etwa sagen: ,wahr ist eine Vorstellung, wenn sie mit der Wirklichkeit [(7)] übereinstimmt', so wäre damit nichts gewonnen, denn, um dies anzuwenden, müsste man [(8)] in einem gegebenen Falle entscheiden, ob eine Vorstellung mit der Wirklichkeit übereinstimme,	If, for example, we wished to say 'an idea is true if it [(7)] agrees with reality' nothing would have been achieved, since in order to apply this definition we should have to decide whether [(8)] some idea or other did agree with reality.	If, for example, we wished to say: 'An idea is true if it [(7)] corresponds to reality', nothing would have been achieved, since in order to apply this definition we should have to decide [(8)] in a particular case whether an idea did correspond to reality,
[(9)] mit anderen Worten: ob es wahr sei, dass die Vorstellung mit der Wirklichkeit übereinstimme.	[(9)] [no translation]	[(9)] or in other words: whether it is true that the idea corresponds to reality.
Es müsste also das Definierte selbst vorausgesetzt werden.	Thus we should have to presuppose the very thing that is being defined.	Thus we should have to presuppose the very thing that is defined.
Dasselbe gälte von jeder Erklärung von dieser Form:	The same would hold of any definition of the form	The same would hold of any definition of the form:
,A ist wahr, wenn es die und die Eigenschaften hat, oder zu dem und dem in der und der Beziehung steht'.	'A is true if and only if it has such-and-such properties or stands in such-and-such a relation to such-and-such a thing'.	'A is true if and only if it has such-and-such properties or stands in such-and-such a relation to such-and-such a thing'.
Immer käme es wieder im gegebenen Falle darauf an,	In each case in hand it would always come back to the question	In each case in hand it would always come back to the question
ob es wahr sei, dass A die und die Eigenschaften habe, zu dem und dem in der und der Beziehung stehe.	whether it is true that A has such-and-such properties, or stands in such-and-such a relation to such-and-such a thing.	whether it is true that A has such-and-such properties, or stands in such-and-such a relation to such-and-such a thing.
Wahrheit ist offenbar etwas so Ursprüngliches und Einfaches, dass eine Zurückführung auf noch Einfacheres nicht möglich ist.	Truth is obviously something so primitive and simple that it is not possible to reduce it to anything still simpler.	Truth is obviously something so primitive and simple that it is not possible to reduce it to anything still simpler.

A.3   Some remarks on the translations

'Der Gedanke':

(1) 'das Gebiet, wo überhaupt Wahrheit in Frage kommen könne'.

The best translation of this sentence that we have found is as follows:

'the region in which the question "Is it true?" could be in principle applicable'.

Long and White translate a parallel formulation in Frege's '[Notes for Ludwig Darmstaedter]' analogously (see Frege, 1919, p. 273 (253); see the Frege quote, contained in the Sluga quote in section 1.1). We have adopted their translation both in the third and in the fourth paragraphs of 'Der Gedanke'.

(2) 'Oder doch?'

In the German original, Frege here raises the question 'Or does it?' ('Oder doch?') immediately following the last sentence of the second argument ('Truth does not admit of more or less'), thus establishing a *close link* between the third and the second arguments. In the translation by Geach and Stoothoff this link has considerably been weakened. Frege's question 'Oder doch?' is not even translated as an independent question but used added at the beginning of the next sentence as 'but'. This translation is inadequate because no one would retranslate 'but' back into German as 'Oder doch?'. In addition Geach and Stoothoff have separated the second and the third arguments by adding a dash after the second argument.

(3) 'Und damit'

'And with that': Geach and Stoothoff's translation renders the words 'und damit', here used by Frege in the German original, as 'and then' in English. This does not quite capture the meaning of 'damit' in this context, which – unlike the English 'then' – emphasizes a logical implication in reference to the inquiry Frege addresses in the previous sentence. In other words, 'und damit' should be read as meaning 'and with that' or 'and thereby', indicating that the very inquiry mentioned in the previous sentence would mean or imply being confronted with a question of the same kind. We therefore render this expression as 'and with that' here, making another change to the standard translation of Frege's text.

(4) 'dieser Versuch, die Wahrheit als eine Übereinstimmung zu erklären'

In this and the next sentence the phrases '*this* attempt' ('*dieser* Versuch' (my italics)) and '*any other* attempt' ('*jeder andere* Versuch' (my italics)) are in juxtaposition to each other. This obvious juxtaposition is lost in Geach's translation 'the attempted explanation'.

(5) '*So* scheitert aber auch jeder andere Versuch'

Geach has here simply 'and' as translation of the German 'so'. This does not sufficiently capture the connotation of analogy in the use of 'so' in the German original. Therefore, it is here rendered as 'likewise'.

(6) 'die Frage nach dem Wahrsein'

Geach/Stoothoff translate 'die Frage nach dem Wahrsein' as 'the question of truth'. They use the same formulation also to translate 'wo Wahrheit in Frage kommt' (cf. note (1)). Yet these are two different grammatical structures in German that have different Fregean senses as well.

'Logik':

(7) 'übereinstimmen mit'

We translate the German phrase 'übereinstimmen mit' in 'Logik' with the same English expression that is standardly used in translations of 'Der Gedanke', namely 'correspond to'.

(8) 'in einem gegebenen Fall'

The German phrase 'in einem gegebenen Fall' is here rendered in English as 'in a particular case'; the Long/White translation uses 'some idea or other' to capture the meaning of the phrase, but this rendering does not sufficiently capture the significance of Frege's emphasis on particular cases of application (which is relevant in the context of section 9.1).

(9) 'mit anderen Worten: ob es wahr sei, dass die Vorstellung mit der Wirklichkeit übereinstimme.'

This part of the sentence is entirely missing from the Long/White translation; it is here directly translated from the German original.

Notes

1 Introduction: In Tarski's Shadow

1. See, however, section 15.4.
2. Frege, *Posthumous Writings*, 253, transl. modified. [Note by Sluga; see Frege, '[Notes for Ludwig Darmstaedter]', p. 362]
3. Dummett, 1973, pp. 442ff; Künne, 1985, pp. 136–8, and 2003, pp. 129–33; Soames, 1999, p. 24ff; and Stuhlmann-Laeisz, 1995, pp. 14–28. Künne has in the meantime slightly modified his assessment of Frege's arguments; in his essay 'Wahrheit' (1985) he still presented all of the Fregean arguments that he discussed (according to my count, arguments 1, 3 and 4) as flawed; in his book *Conceptions of Truth* (2003), pp. 129–33, however, he grants Frege's fourth argument a certain (severely restricted) degree of validity, while still rejecting Frege's claim that this argument establishes the general indefinability of truth; and in his book *Die Philosophische Logik Gottlob Freges* (2010), pp. 391–423, his interpretation is slightly altered again. Other more positive assessments of Frege's arguments have been provided by Greimann, Moreno, Sluga and Stepanians. Cf. Bibliography.

2 The Context: The Question of Truth Bearers

1. See for this the Popper and Davidson quotes in section 1.1.
2. The parenthized numbers (i), (ii), ..., (x) appearing after a Frege quote indicate the corresponding sub-paragraph in Frege's text of the third and fourth paragraphs of 'Der Gedanke' (see section 3.1).
3. See quotations (viii) and (ix) in Chapter 3. There we have two occurrences of the word 'really' and one occurrence of the word 'improper'. See also section 11.5.
4. The difference will be discussed later in section 11.4.

3 Frege's Text and Its Argumentative Structure

1. In the original text, this passage initiates the fourth paragraph.
2. This shows just how important the demarcation of arguments and quotes really is for the interpretation of a text.
3. Since the concept of truth as correspondence can be applied analogously to pictures and ideas (see section 2.6), for reasons of simplicity I shall restrict my discussion of the definitions to the case of ideas.

4 The First Argument: Scientific Truth Is Absolute

1. Since I understand (D_0) as a variation of (D_1), I shall neglect (D_0) in the following and mostly refer to (D_1).
2. See also Stuhlmann-Laeisz, 1995, pp. 17–19.

231

5   The Second Argument: Scientific Truth Is Perfect

1. Strictly speaking, sentence (5) does not contain the adjective 'perfect' but the adverb 'perfectly'. Though the word 'perfect' occurs as an attributive adjective in phrases such as 'a perfect truth' and 'a perfect correspondence', it becomes an adverbial of manner in 'x is perfectly true' and 'x perfectly corresponds to f(x)'.

6   The Third Argument: Scientific Truth Is Independent

1. Dividing up each of sentences (G4), (G5), and (G6) into an A and a B part makes it easier to refer to the respective sentence parts (G4a), (G4b), and so on in later discussions. The sentences as a whole shall be referenced as (G4), (G5), and so on.
2. I use the symbol '*' in '(G5*)' (and similar expressions) whenever I present a variant, in my own formulation, of Frege's original sentences.
3. According to Frege, *any* question about whether something is I-true can be answered only via the corresponding question about whether it is S-true that.... There are, however, questions about S-truth that do not require a corresponding question about I-truth. When we ask, for example, whether 7 + 5 = 12 we do not refer to ideas; rather, we wish to know whether a specific thought (sentence) is true. In short: an application of the concept of I-truth always depends on an application of the concept of S-truth, but not vice versa.
4. I distinguish here explicitly between the *expression* 'P' (in single quotation marks) and the *concept P* (in italics). To the extent that definition 'P:= A and B' is correct the expressions 'P' and 'A and B' are distinct while the concepts *P* and *A and B* are identical.
5. My emphasis. See also the complete quote of this parallel passage in section 7.1.

7   Parallels in Frege's 'Logik'

1. You will find the two parallel passages quoted in full a little later in this section.
2. See for this also sections 6.1 and 6.8. See Greimann, 1994, p. 78; Künne, 1985, p. 137.
3. On the broad and the narrow reading of 'characteristic' see section 7.3.
4. Of course, (L9) and (G9) themselves are not proper definitions; rather (L9) is merely a definitional scheme, while (G9) constitutes something like a structural description of definitions. I hope that my abbreviating manner of speaking does not lead to misunderstandings here.
5. Recall my thesis that Frege abandons the circle objection in the G version of the third argument because he has come to regard it as invalid (see section 7.5). Consequently, he has to make sure that the third definition in G is not a special case of the fourth one, since he does uphold the circle objection in the fourth argument.
6. I consider this a confirmation of my reading of the third argument. However, I do not see how the distinctness of the truth concepts in (G5b) could explain a transition from a circle to a regress.

7. With the sentence 'Like any other (!) picture, an idea is not true in itself [...]' Frege makes it clear that he considers ideas to be a sort of pictures. That is, ideas themselves *are* pictures (my note).

8. Frege, 1897, pp. 145–6 (133–4); my emphasis; the structuring of the quoted text into three paragraphs and their line breaks is my doing. In Frege's original text the first two paragraphs are a component of *one* rather long paragraph, and only the third paragraph containing the 'treadmill' is separated from the others.

9. With respect to its subject matter, this argument strongly brings to mind the following argument by Kant: 'Truth, it is said, consists in the agreement of cognition with its object. In consequence of this mere nominal explanation, my cognition, to count as true, is supposed to agree with its object. Now I can compare the object with my cognition, however, only *by cognizing it*. Hence my cognition is supposed to confirm itself, which is far short of being sufficient for truth. For since the object is outside me, the cognition in me, all I can ever pass judgment on is whether my cognition of the object agrees with my cognition of the object' (Kant, 1800, pp. 557–8).

10. Künne, 2003, p. 129. Künne uses brackets instead of parentheses. I am using parentheses inside the quotation as I prefer to reserve brackets for comments. While Künne here (p. 129) appears to identify the treadmill argument with the third argument in 'Der Gedanke', on p. 132 he reads it as a paraphrase of the fourth argument. Greimann reads the third argument in 'Der Gedanke' as a regress argument and finds that this same regress is described 'by Frege in the clearest and most conspicuous manner [...] in [the treadmill argument]' (Greimann, 2003, pp. 203–4).

8 The Fourth Argument: The Circle Objection

1. Dummett, 1973, pp. 442–3, p. 451; Künne, 1985, p. 137; Soames, 1999, pp. 24–5; Stuhlmann-Laeisz, 1995, pp. 25–6.

2. Greimann, 2003, p. 207. Greimann uses brackets [...] instead of parentheses (...) and external brackets [(...)] instead of double parentheses ((...)). I am using parentheses inside the quotation as I prefer to reserve brackets for comments.

3. Cf. Moreno, 1996, p. 33; Stepanians, 1998, pp. 86–7.

4. This restriction to S-truth is necessary, as has been shown in section 7.6.

5. Consider as an example 'x is a thought of God (of Kant, of Frege, of Quine).' Such criteria of truth, if you wish to call them this, have always been extremely popular. If 'reality' is a singular term then this structure is present also in the definition presented as a counterexample by Künne: 'x is true:= x agrees with reality.'

9 The Omnipresence of Truth

1. Lotter, 2006, p. 431. Lotter then continues: 'However, it is perhaps not even necessary for a good and useful interpretation to have captured the exact intentions of an author if the interpretation is understood as a means to show the depth inherent in the author's thought (whether or not he was himself entirely clear about every detail). This, in any case, strikes me as both

more constructive and more interesting than trying to prove that Frege – who, after all, was one of the most important and creative logicians of the 19th and 20th century – was making simple beginner's mistakes when arguing about the concept of truth, as had been common in much of mainstream Frege literature in the 20th century.'

10 Dummett's Regress

1. Dummett, 1973, p. 442f. In Dummett's original text the following four quotations are immediately subsequent to each other. It becomes obvious on pages 442 and 445 that Dummett's exposition of Frege's arguments refers to the version in 'Der Gedanke'.

12 The Fifth Argument: Frege's Regress

1. Rather than placing multiple quotation marks one after the other I prefer to use the indices '0', '1', '2',..., inside the quotation marks, which makes the text more readable.
2. This objection was raised by an anonymous reviewer of an earlier version of this book whom I will call reviewer B.

13 The Road to a Novel Interpretation

1. The translation follows the modified standard translation used in this book. Künne's Frege quote and its translation are as follows: 'Eine Übereinstimmung ist eine Beziehung. Dem widerspricht aber die Gebrauchsweise des Wortes "wahr", das kein Beziehungswort ist. [!] (An agreement is a relation. But this is incompatible with the use of the word "true", which is not a relation word. [!])' (Künne, 2003, p. 93; Künne uses brackets instead of parentheses). This is another Frege quote that Künne altered without highlighting the changes: Künne considerably shortened the second sentence of the quoted passage omitting the passage 'does not contain any indication of anything else to which something is to correspond'. But is this passage irrelevant for the argument?
2. Frege's German original does not use the word 'question' ('Frage') but another word ('Wunsch') that is normally translated as 'desire'. Yet the desire to demarcate the region of truth bearers implies the question of how this region can be appropriately demarcated.
3. My indexing of the various occurrences of 'true' as 'true$_1$' and 'true$_2$' serves exclusively the purpose of easier reference to these occurrences.

14 Absolute or Relative Truth?

1. Analogously Frege writes: 'All the same, it is something worth thinking about that we cannot recognize a property of a thing without at the same time finding the thought *this thing has this property* to be true. So with every

property of a thing there is tied up a property of a thought, namely truth'
(Frege, 1918a, p. 61; Geach's emphasis).
2. This sentence is a condensed version of the following sentence: 'The thought
which is expressed by the English sentence in line (*) is not true.'
3. On the notion of relative truth see section 2.3.
4. Quine abbreviates 'true in S_0', 'true in S_1' and so on to 'true$_0$', 'true$_1$' and so
on, that is, he uses a truth predicate with subscripts.

Bibliography

> A young Greek boasted of having devoured an infinite number of books. Aristippus replied: 'The healthiest are not those that eat much, but those that digest well.'

Beaney, M. (1997) *The Frege Reader*, 9th edn (Oxford: Blackwell, 2006).

Candlish, S. (2007) *The Russell/Bradley Dispute and its Significance for Twentieth-Century Philosophy* (Basingstoke and New York: Palgrave Macmillan).

Carnap, R. (1931) 'Überwindung der Metaphysik durch logische Analyse der Sprache' in *Erkenntnis*, II, 219–41. Trans. A. Pap as 'The Elimination of Metaphysics through the Logical Analysis of Language', in A. J. Ayer (ed.) *Logical Positivism* (New York: The Free Press, 1959).

Davidson, D. (1969) 'True to the Facts', *Journal of Philosophy*, LXVI, 748–764.

Dummett, M. (1973) *Frege: Philosophy of Language*, 2nd edn (Cambridge, MA: Harvard University Press, 1981).

Frege, G. (1884) *Die Grundlagen der Arithmetik*. In excerpts trans. M. Beaney as *The Foundations of Arithmetic*, in Beaney, 2006.

—— (1892a) 'Über Sinn und Bedeutung', in Frege, 1967. Trans. M. Black as 'On Sense and Meaning', in Frege, 1984.

—— (1892b) '[Ausführungen über Sinn und Bedeutung]', in Frege, 1969. Trans. P. Long and R. White as '[Comments on Sense and Meaning]' in Frege, 1979.

—— (1893) *Grundgesetze der Arithmetik*, Band I. In excerpts trans. M. Beaney as *Grundgesetze der Arithmetik*, Volume I, in Beaney, 2006.

—— (1897) 'Logik', in Frege, 1969. Trans. P. Long and R. White as 'Logic', in Frege, 1979.

—— (1914) 'Logik in der Mathematik', in Frege, 1969. Trans. P. Long and R. White as 'Logic in Mathematics' in Frege, 1979.

—— (1918a) 'Der Gedanke', in Frege, 1967. Trans. P. Geach and R. H. Stoothoff as 'Thought', in Frege, 1984.

—— (1918b) 'Die Verneinung', in Frege, 1967. Trans. P. Geach as 'Negation' in Frege, 1984.

—— (1919) '[Aufzeichnungen für Ludwig Darmstaedter]' in Frege, 1969. Trans. P. Long and R. White as '[Notes for Ludwig Darmstaedter]' in Frege, 1979.

—— (1966) *Logische Untersuchungen*, ed. G. Patzig, 5th edn (Göttingen: Vandenhoeck & Ruprecht, 2003).

—— (1967) *Kleine Schriften*, ed. I. Angelelli, 2nd edn (Hildesheim: Georg Olms, 1990).

—— (1969) *Nachgelassene Schriften*, ed. H. Hermes, F. Kambartel, F. Kaulbach, 2nd edn (Hamburg: Felix Meiner, 1983).

—— (1979) *Posthumous Writings*, trans. P. Long and R. White, ed. H. Hermes, F. Kambartel, F. Kaulbach (Oxford: Basil Blackwell).

—— (1984), *Collected Papers on Mathematics, Logic, and Philosophy*, ed. B. McGuinness (Oxford: Basil Blackwell).

Greimann, D. (1994) 'Freges These der Undefinierbarkeit von Wahrheit. Eine Rekonstruktion ihres Inhalts und ihrer Begründung', *Grazer Philosophische Studien*, XLVII, 77–114.

—— (2003) *Freges Konzeption der Wahrheit* (Hildesheim et al.: Georg Olms Verlag).

Hartmann, N. (1926) *Ethik*, 4th edn (Berlin: Walter de Gruyter, 1962).

Heidegger, M. (1929) *Was ist Metaphysik?*, 15th edn (Frankfurt a. M.: Klostermann, 1998).

Kant, I. (1785) *Grundlegung zur Metaphysik der Sitten*. Trans. M. Gregor as *Groundwork of the Metaphysics of Morals*, in Kant, 1999.

—— (1788) *Kritik der praktischen Vernunft*. Trans. M. Gregor as *Critique of Practical Reason*, in Kant, 1999.

—— (1790) *Kritik der Urteilskraft*. Trans. P. Guyer and E. Matthews as *Critique of the Power of Judgment*, ed. P. Guyer (Cambridge: Cambridge University Press, 2001).

—— (1800) *Logik*. Trans. J. M. Young as *The Jäsche Logic*, in Kant, 1992.

—— (1992) *Lectures on Logic*. Trans. and ed. J. M. Young (Cambridge: Cambridge University Press).

—— (1999) *Practical Philosophy*, trans. and ed. M. Gregor (Cambridge: Cambridge University Press).

Künne, W. (1985) 'Wahrheit' in E. Martens and H. Schnädelbach (eds) *Philosophie. Ein Grundkurs* (Reinbek: Rowohlt Taschenbuch).

—— (2003) *Conceptions of Truth* (Oxford: Oxford University Press).

—— (2010) *Die Philosophische Logik Gottlob Freges* (Frankfurt a. M.: Klostermann).

Kutschera, F.v. (1989) *Gottlob Frege. Eine Einführung in sein Werk* (Berlin et al.: Walter de Gruyter).

Lotter, D. (2006) 'Review of Ulrich Pardey: Freges Kritik an der Korrespondenztheorie der Wahrheit. Eine Verteidigung gegen die Einwände von Dummett, Künne, Soames und Stuhlmann-Laeisz', Journal for General Philosophy of Science, XXXVII, II, 425–36.

Mendelsohn, R.L. (2005) *The Philosophy of Gottlob Frege* (Cambridge: Cambridge University Press).

Moreno, L.F. (1996) 'Die Undefinierbarkeit der Wahrheit bei Frege', *Dialectica*, L, I, 25–35.

Pardey, U. (1994) *Identität, Existenz und Reflexivität* (Weinheim: Beltz Athenäum).

—— (2004) *Freges Kritik an der Korrespondenztheorie der Wahrheit. Eine Verteidigung gegen die Einwände von Dummett, Künne, Soames und Stuhlmann-Laeisz* (Paderborn: Mentis Verlag).

—— (2006) *Begriffskonflikte in Sprache, Logik, Metaphysik* (Paderborn: Mentis Verlag).

Patzig, G. (1966) 'Einleitung' in Frege, 1966.

Popper, K.R. (1963) *Conjectures and Refutations: The Growth of Scientific Knowledge* (London: Routledge and Kegan Paul)

Quine, W.V.O. (1966) 'The Ways of Paradox', in W. V. O. Quine (ed.) *The Ways of Paradox and other Essays*, revised and enlarged edition (Cambridge, MA: Harvard University Press, 1994).

Sluga, H. (1999) 'Truth before Tarski' in J. Woleński and E. Köhler (eds) *Alfred Tarski and the Vienna Circle. Austro-Polish Connections in Logical Empiricism* (Dordrecht et al.: Kluwer Academish Publishers).

Soames, S. (1999) *Understanding Truth* (New York et al.: Oxford University Press).

Stepanians, M. (1998) *Frege und Husserl über Urteilen und Denken* (Paderborn et al.: Ferdinand Schöningh).

—— (2001) *Gottlob Frege zur Einführung* (Hamburg: Junius).

Stuhlmann-Laeisz, R. (1995) *Gottlob Freges 'Logische Untersuchungen'* (Darmstadt: Wissenschaftliche Buchgesellschaft).

Tarski, A. (1935) 'The Concept of Truth in Formalized Languages', in A. Tarski (ed.) *Logic, Semantics, Metamathematics*, 2nd edn by J. Corcoran (Indianapolis: Hacket Publishing Company, 1983).

—— (1944) 'The Semantic Conception of Truth' *Philosophy and Phenomenological Research*, IV, 341–76. Repr.: A. Tarski (1952) 'The Semantic Conception of Truth' in L. Linsky (ed) *Semantics and the Philosophy of Language* (Urbana et al.: University of Illinois Press).

Index

Printed in the United States
By Bookmasters

Printed in the United States
By Bookmasters